INSIGHTS

Africa's future ...
can biosciences
contribute?

**Insights: Africa's future ...
can biosciences contribute?**

Essays compiled by Brian Heap[1,2]
and David Bennett[1]

1   St Edmund's College, Cambridge CB3 0BN
2   Centre of Development Studies, Alison Richard Building,
    7 West Road, Cambridge CB3 9DT

Published in 2013 by Banson, 17f Sturton Street, Cambridge CB1 2SN

ISBN (paperback) 978-0-9563387-5-4
        (hardback)   978-0-9563387-6-1

Citation: Heap, R.B. and Bennett, D.J. (eds) 2013. *Insights: Africa's Future
... Can Biosciences Contribute?* Banson/B4FA.

This publication was made possible through the support of grants from
the John Templeton Foundation and the Malaysian Commonwealth
Studies Centre, c/o Trinity College, Cambridge CB2 1TQ.

Copy editor: Anna Hodson
Design and layout: Banson

Printed in the UK by Lavenham Press

Africa's future ...
can biosciences
contribute?

... the greatest service which can be rendered any country is to add an useful plant to its culture; especially a bread grain.

Thomas Jefferson (American Founding Father)
Memorandum of Services to My Country
after 2 September 1800

Our Academy concluded that recently established methods of preparing transgenic organisms follow natural laws of biological evolution and bear no risks anchored in the methodology of genetic engineering ... The beneficial prospects for improving widely used nutritional crops can be expected to alleviate the still existing malnutrition and hunger in the human population of the developing world.

Werner Arber (Nobel Laureate 1978, Physiology or Medicine)
President, Pontifical Academy of Sciences
at the General Assembly of the Synod of Bishops,
Vatican City, October 2012

Biotechnology is a tool rather than an end in itself, but in some contexts it is the most appropriate tool.

Kanayo F. Nwanze
President, International Fund for Agricultural Development
at McGill University, Montreal, Canada,
1 November 2012

... sustaining African economic prosperity will require significant efforts to modernize the continent's economy through the application of science and technology in agriculture.

Calestous Juma

The New Harvest: Agricultural Innovation in Africa

Oxford University Press, 2011

If you care about the poorest, you care about agriculture. Investments in agriculture are the best weapons against hunger and poverty, and they have made life better for billions of people. The international agriculture community needs to be more innovative, coordinated, and focused to help poor farmers grow more. If we can do that, we can dramatically reduce suffering and build self-sufficiency.

Bill Gates

at the International Fund for Agricultural Development, 2012

We can feed the world but it will not be easy. There are no magic bullets. The answer lies in seeking win–win–win solutions where there are economic, social and environmental benefits.

Gordon Conway

One Billion Hungry: Can We Feed the World?

Comstock Publishing Associates, 2012

Preface

Sir John Templeton had an enduring interest in the Big Questions of life and what he referred to as ultimate reality – *Do we have free will? Is information the basis of reality? Will machines ever become human?* He was also fascinated by genetics and its potential to improve life for humanity, as it might for the 1 billion people who are currently undernourished. And then there is the predicted increase from 7 to 9 billion people living on this planet by 2050 and the need for even more food produced in a sustainable way – a grand challenge for the best brains and entrepreneurs alike, whether in laboratories, farms, businesses or partnerships.

The Reverend Thomas Malthus was a brilliant mathematician. Writing his famous essay over 200 years ago while at Jesus College, Cambridge he reasoned that population growth would exceed our ability to feed ourselves. He predicted dire social unrest. He could not have foreseen the ingenuity of plant breeders and geneticists who by cross-breeding and selection developed higher-yielding crops which over the years kept pace with the burgeoning increase in global population. Not everyone benefited but the effects of global famine have been averted so far. Year on year, however, the increase in yields is now slowing in the developed world and today's science and technology are even more important for best agronomic practices if we are to find a way to sustainable food production.

Insights was commissioned as a key part of a grant awarded by the Templeton Foundation to examine the implementation of biosciences for farming in Africa (www.B4FA.org). The essays are eclectic and personal, they take the style of other Templeton essays in being sharply focused, and they are meant to inform decision-makers whether relaxing on long-haul flights, or in deepest deliberation with colleagues. They do not advocate a position. They argue from experience, and offer an authoritative, independent and peer-reviewed brief.

The collection shows that the grand challenge of the *rights of all to food*, highlighted by Phil Bloomer of Oxfam, is being addressed *by Africans for Africa*, as Calestous Juma of Harvard shows in the first essay. Of course, it is a long and difficult way ahead. Africa is a continent of 54 countries where the large majority of farmers are smallholders and a crucial challenge is to improve low crop yields by empowering the farmers themselves. 'Women farmers are the pillars of African agriculture', says Lindiwe Majele Sibanda, CEO of the Food, Agriculture and Natural Resources Policy Analysis Network (FANRPAN), and over two-thirds of all women in Africa are employed in the agricultural sector and produce nearly 90 per cent of food on the continent. They are responsible for growing, selling, buying and preparing food for their families.

African countries are currently among the fastest growing economies in the world and Africa's readiness for transformation is enriched by its natural resources, whether land and minerals, solar radiation in North Africa and the Sahara, the huge potential for food production in the central and eastern regions and Southern Africa, or the unrivalled diversity of environments and people. Essays speak of the crucial role of high-quality seeds for smallholder farmers. How certain African countries became early adopters of new technologies, and how industry responded to new opportunities. How the skills of the business world can be combined with public initiatives for mutual benefit. How world-class research can be developed locally, and train the next generation of African plant breeders. Why African youth holds the key to the future and needs to be inspired about the role of biosciences in African farming.

How water can be delivered to places where it is life-promoting. And a glimpse of the very latest prospects for plant protection.

In the past people were fearful of introducing potatoes into the Western diet for what now seem extraordinary reasons – they were not to be found in the Bible. Recent advances as well as exploitation of plant breeding have also evoked concern and antagonism. So there are insights into simple but profound ethical questions about the new technologies: Are they safe? Are they fair? Are they natural? Are they needed in the face of climate change? Do they need to be protected? Are they designed to benefit the consumer?

We hope these *Insights*, alongside our media fellowship programme and associated activities (www.B4FA.org), will inspire and inform political and academic leaders, enlighten evidence-based policy-making, and encourage balanced debates in schools, colleges, universities, religious centres, villages and the media. We also hope they will assist measured analysis and communication by all who engage in education, whatever the level.

Above all, we hope you enjoy these reflections and are moved by them – we have been.

Brian Heap   **David Bennett**

Contents

Biosciences in Africa's economic transformation

Jorgen Schytte/Still Pics/Specialist Stock

Calestous Juma

Africa countries are currently among the fastest growing economies in the world. Much of this is linked to China's commodity demand, which is expected to decline in coming years as China's growth slows down.

External stimuli, however, have tended to mask endogenous sources of growth which are driven by investments in agriculture. Such endogenous growth will rely heavily on Africa's investment in the life sciences, with genetics serving both as foundational field and as source of metaphors on how to understand the world.

The continent's renewed interest in fostering agricultural development has come at a critical moment when the impacts of climate change are becoming

Africa's economic transformation is starting in the new age of biology

11

more discernible and the need to rethink Africa's agricultural strategies accordingly is more pressing.

African agriculture will need to intensify the use of science and technology more than would have been the case without the threats of climate change. Investments in science and technology will be required along the entire agricultural value chain from resource intelligence through production, marketing, storage and ecological rehabilitation.

Science, technology and economy

Agriculture will continue to play a significant role in Africa's economic transformation. This is for two main reasons. First, agriculture accounts for 32 per cent of Africa's GDP and for nearly 65 per cent of employment. In effect, it will not be possible to promote prosperity in Africa without significant focus on agricultural transformation.

Second, so far much of the discussion on the role of science and technology in economic growth has tended to focus on high-tech, urban-based industries. The application of science and technology to Africa's economic transformation will need to focus on agriculture which is still the base upon which much of the growth and employment are founded. This does not mean that technological innovation will have to start in the agricultural sector. Given Africa's economic structure new technologies will quickly find agricultural applications irrespective of their origins. The mobile money revolution that started in Kenya is a good example. Mobile phones first took root in urban areas and spread to rural areas. But it was the need to find a low-cost way of enabling rural women to repay micro-

... it will not be possible to promote prosperity in Africa without significant focus on agricultural transformation

loans that inspired the creation of the mobile money (M-Pesa) revolution that has now gone global.

Technologies that focus on the needs of women are likely to have the widest adoption rates and impact. This is mainly because of the dominant role that women play in African economies. It is also for this reason that efforts to advance science, technology and engineering education in Africa need to specifically focus on women. Analogues of the mobile money revolution are already being piloted in health, education and energy in rural Africa.

The century of Africa

Mobile communication will create a wide range of innovative ventures for Africa. But the technological transformation of rural Africa will require foundational investments in basic infrastructure such as energy, transportation, water supply and irrigation. It is easy to use a mobile phone to find out where to get the best price for one's produce. But one has to move the produce there, which requires a reliable transportation network. It is estimated that it will take nearly US$93 billion a year over the next decade to meet Africa's infrastructure needs.

Nearly 60 per cent of the world's available arable land is in Africa. Part of this land is being leased out to hedge funds with little agricultural know-how. Careful design and implementation of this critical agricultural asset will allow Africa to feed itself as well as to export food to the rest of the world. Doing so in a smart way will require using emerging technologies that maximise sustainable intensification.

The century of biology

Timing defines nations. South Asian economies embarked on their growth path at the dawn of the microelectronics revolution. Africa's economic transformation is starting in the new age of biology – both as a field of scientific endeavour and as a metaphor on how we view the world. The world of genetics captures both phenomena and will most likely offer Africa its opportunity to become an important player in the global knowledge ecology.

There are already indications of this. Young African scientists, many of them women, are making significant contributions to medical and agricultural research using genetics and foundational knowledge. They are designing new diagnostics for human and animal diseases and developing new crop varieties that tolerate drought, resist pests and outcompete weeds.

Governing the future

Africa has so far suffered greatly from a lack of high-level leadership that appreciates the importance of innovation in development. Efforts to graft Western democracy on to Africa have had mixed results, with too much focus on elections and too little attention paid to building democratic institutions such as party platforms, think-tanks and succession mechanisms within parties.

Africa appears to be undergoing a rapid jump from autocracy to technocracy without really having built what one can honestly call democracies. In 2005 Namibia's founding President Sam Nujoma stepped down as head of state and registered as a Masters student in geology at the University of Namibia where he was also Chancellor. He graduated in 2009.

In the past only a handful of African presidents had any training in the sciences. Today Angola, Egypt, Eritrea, Ethiopia, Nigeria, Senegal, Somalia and Tunisia have engineers or medics as heads of state. The appeal to technocracy might just be a way to go around the ethnic patronage that underpins much of Africa's corruption, nepotism and misgovernment.

But there is more to this seemingly accidentally rising technocracy in Africa. These are leaders who will readily appreciate the importance of science and technology in economic transformation. Many of them have risen to positions of leadership because their followers

... young African scientists, many of them women, are making significant contributions to medical and agricultural research

value their pragmatism. Their technical training has prepared them to understand the evolutionary character of economies.

This is indeed the century of biology, with genetics serving as its disciplinary and metaphorical paradigm. It is going to be African leaders who are either trained in the scientific, technological and engineering fields or leaders who appreciate the dynamics of change through time that will turn this vision into reality.

Further reading
Juma C. (2011) *The New Harvest: Agricultural Innovation in Africa.* Oxford, UK: Oxford University Press.

Professor Calestous Juma FRS is Professor of the Practice of International Development and Director of the Science, Technology, and Globalisation Project at Harvard University. Mailbox 53, 79 JFK Street, Cambridge, MA 02138, USA. calestous_juma@harvard.edu

The right to food in a changing world

Smandy/Dreamstime.com

Phil Bloomer

Tonight, around 870 million people will go to bed hungry: one in eight of us. More than half will be in Asia, many will be in Africa. A majority will be women and girls. Most will go to bed hungry not because there is not enough food in the world, or even a local food shortage, but simply because they are too poor to buy the food that is available. Feeding the 870 million hungry in 2013 is therefore not principally a technical challenge for agriculture, but rather a political challenge to governments to end hunger by addressing poverty and vulnerability. This is possible. Under President Lula, the number of hungry people in Brazil dropped by 40 per cent in just ten years through a modest redistribution of wealth. Vietnam halved poverty in nine years through focusing agricultural investment where poor and hungry people are.

There is more than enough food for everyone on the planet: one-third of the world's food supply is lost post-harvest. However, as the UK government's Chief

Scientific Adviser, Sir John Beddington has pointed out, we are now entering a new era of global food insecurity, unless radical action is taken. After 30 years of declining relative food prices we appear to be entering a long period of rising food prices and price volatility. A 'Perfect Storm'[1] of interlocking factors is threatening food production and tightening food markets in the long term, including: climate change impacts, absurd biofuel policies, rising oil prices, dietary upgrading in Asia, land grabs, speculation in commodity markets, and population growth to around 9 billion by 2050. These factors are not only driving food price increases but also creating price volatility. We are now in a third food price spike in just four years. Poor people are trapped by their income poverty, but also by the increasing numbers of shocks, caused by food price hikes and extreme weather, that lead to a catastrophic loss of their assets that can take a decade to recover.

One key solution to this threat of worsening hunger in Africa and other continents is to increase investment in poor farmers, especially women

One key solution to this threat of worsening hunger in Africa and other continents is to increase investment in poor farmers, especially women. This makes sense both morally and economically. It is the 500 million smallholder farms that provide livelihoods for the poorest 2 billion people on this planet; the same 2 billion that are generally ignored by commercial farming and food companies as too poor to bring a return on investment. It is also the same 500 million smallholder farms where some of the greatest productivity gains can be achieved in terms of yield per unit area and per unit input.[2] Investing in them would create enormous potential gains in food production and food security for poorer countries and also translate directly into a more equitable and effective economic growth. A 1 per cent increase in per capita GDP in agriculture

17

... the default of agricultural biotechnologies has been to support large-scale export crops rather than address food insecurity

reduces the depth of poverty at least five times more than a similar increase outside agriculture.[3]

Bioscience has a role to play as we face up to the challenge of global food insecurity. But it is no magic bullet. Most productivity gains in African smallholdings will come from enhanced extension services, investment for irrigation, low-input crop husbandry such as inter-cropping, fairer markets, market information through mobile phones, and hard and soft infrastructure from roads to marketing boards. This is where the bulk of the investment must go to enhance food security. But increased productivity also depends on improved seed and better varieties of staple and Africa-specific crops (cassava, banana, sweet potato, etc.) produced by marker-assisted conventional breeding, vegetative propagation and, where appropriate, GM technologies.

Dr Mohamed 'Mo' Ibrahim, UK's mobile communications entrepreneur and billionaire,[4] has pointed out repeatedly that we need African governments to set national agricultural policies based on delivering food security through directing public and private investment to where it can bring the greatest economic and social benefit for the poor. An example of this is the current strategic plan of West and Central African Countries (CORAF/WECARD),[5] which identifies food security as the priority and sets out a research agenda for staple food crops such as sorghum, maize, rice and cassava alongside export crops like cotton and cocoa.

We also need international support to help achieve the promise of large productivity gains in smallholder agriculture: from the Brazilian research institute, Embrapa's support to Ghana for smallholder cowpea yield increase through rhizobium inoculants, to the UK's research institute at Rothamsted 'push–pull' low-input cropping system in Kenya that doubles maize yields. None of these involve GM crops.

Nevertheless, there is real potential in accelerated genome improvement of poor people's crops. However, there are two major questions to answer: Where will the investment in plant breeding come from? And if it does come, who will control it and for whose benefit?

Investment in plant breeding (non-GM and GM) for the orphan crops of the poor demands public investment. Agribusiness companies will invest in en-hancing the attributes of the crops needed by commercial agriculture in Africa where a return on investment is guaranteed, but there is little or no commercial incentive for breeding poor people's crops. Up to now, the default of agricul-tural biotechnologies has been to support large-scale export crops rather than address food insecurity.[6] This is the principal reason there are such paltry advances in genome-based yield increases in millet, cowpea and other crops of the poor. Rising economic growth in Africa, and government commitments to investment in agriculture, like the Maputo Declaration, should allow greater investment by the public sector, potentially greater public–private partnerships for accelerated crop breeding (including GM), capacity-building and retention of African scientists, coupled with long-term sustainable business investment.

However, the greatest controversy for GM technologies in Africa rages around the control of agricultural biotechnologies. The domination of whole agricultural systems by

... the greatest controversy for GM technologies in Africa rages around the control of agricultural biotechnologies

19

agribusiness driven partially through the control of patents on crops is clearly not going to enhance productivity gains and income for smallholders in Africa. This concern, and the perceived threat of indentured servitude of smallholders to agribusiness, is what drives most of the opposition to GM both within and outside the continent. This agribusiness model may have brought benefits to large-scale commercial agriculture in the USA (though benefits to the wider society are increasingly questioned and the model is now evolving[7]), and might increase export growth from Africa, but there is no convincing evidence that it holds promise for the food security of the poorest 2 billion on our planet.

The inappropriate agribusiness model must be distinguished from the potential of biotechnologies to help accelerate the breeding of enhanced crop varieties for the poor. But this will require the removal of the stranglehold of counter-productive global intellectual property rules for poor people's crops. That is why we need a renewed democratic debate on biotechnology and intellectual property in the next decade. The debate could be usefully informed by the battle over the last 20 years in the pharmaceutical sector regarding poor people's access to life-saving medicines, orphan diseases of the poor, and the role of technology transfer and generic medicine companies. The appalling mis-calculations of big pharma in defending universal patents at the expense of the lives of poor HIV-AIDS sufferers for a decade, and the rise to power of Brazil, China, India and South Africa have fundamentally changed the in-dustry, and similar forces are already changing agricultural research and markets. Similar lessons, alongside bold public–private investment in

The inappropriate agribusiness model must be distinguished from the potential of biotechnologies to help accelerate the breeding of enhanced crop varieties for the poor

agricultural research, will be necessary for biosciences to play its full role in dealing with the challenges of the 'Perfect Storm' and upholding poor people's right to food.

References

1   *A Perfect Storm* (2009) www.bis.gov.uk/assets/goscience/docs/p/perfect-storm-paper.pdf
2   Conway C., Wilson, K. (2012) *One Billion Hungry*. Ithaca, NY: Cornell University Press.
3   *World Development Report* (2008) New York: The World Bank.
4   www.moibrahimfoundation.org
5   *CORAF/WECARD Strategic Plan 2007–2016* (2007) www.coraf.org/documents/StrategicPlan07_016.pdf
6   Fukuda-Parr S., Orr, A. (2012) *GM Food Crops for Food Security in Africa,* UNDP Working Paper 2012-018. www.undp.org/africa/knowledge/WP-2012-018
7   Reganold J.P., Jackson-Smith D., Batie S.S., *et al.* (2011) Transforming US agriculture. *Science* **332**: 670–1.

Phil Bloomer *is Director of the Campaigns and Policy Division at Oxfam GB.*
pbloomer@oxfam.org.uk

Seed: hope for smallholder farmers?

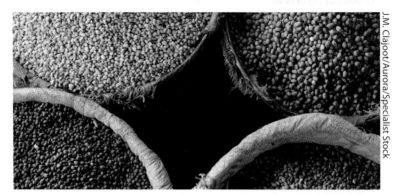

Joe DeVries

There is an interesting paradox at the centre of farming: it is that an activity – indeed, an industry – which involves so much bulk and heavy lifting should be controlled to such an extent by a thing as tiny and delicate as the seed that is planted into the soil. Yet it always has been this way, as we learned when hybrid maize genetics and the breeding of semi-dwarf grain crops permanently altered the physical and social landscape of the American Mid-West, Mexico, India and Asia in the wake of the introduction of revolutionary types of crop varieties in the 1930s, 1950s and 1960s. Seed is the ultimate reach-through technology because its effects in the future extend far beyond immediate applications and intentions.

Africa's farmers are as alive as any in the world to the paradoxical magic of good seed. After more than a quarter of a century of trying to deliver improved seed and better crop management practices to farmers in Africa there is still, for me, nothing equal to the thrill of seeing a smallholder farmer reaching out to receive a new batch of improved seed. That moment always seems to bring a sparkle of hope to the eyes of even the most downtrodden farmer.

I can still recall vividly the days of war in Mozambique when we were distributing 'emergency seed' to farmers affected by the fighting there. The farmers would line up for hours, often in the rains of the new planting season, some of them clothed in tatters and some of them wrapped only in pounded tree-bark, since their clothes had long since rotted away after years of being trapped in the bush. But the gleam in their eyes when they walked away with the seed packs we were distributing always betrayed them: somehow there was hope within the despair brought on by unspeakable hardship. These were people who had lived and died by the viability of the seed all their lives. They understood the magic that could be embodied within a seed, and they had new seed! Sure, the rebels could still push them off the land before they could harvest. They might even wait until the harvest was ready, and then attack. They might burn the crop just as it was drying, along with their homes. But maybe not! Maybe this time the rains would come and the crops would flourish and the forces that had been tearing the country apart from the inside out would go somewhere else. For the moment, there was hope. They had seed. They would plant. New hope for better life would sprout along with the green shoots.

Today things are better. Today Africa's farmers, by and large, are no longer standing in lines to receive free seed. Today, more often than not I see that sparkle in the eyes of exiting customers as I'm entering an agro-dealer's shop where certified seed of any one of the hundreds of new, higher-yielding crop varieties

... there is still nothing equal to the thrill of seeing a smallholder farmer receiving a new batch of improved seed

23

Africa's demands on the new seed would be very different from those that worked wonders in Latin America or Asia

developed in the past few years is being sold. Africa's farmers, along with thousands of African seed entrepreneurs, have discovered the value of improved higher-yielding seed.

To get to this point has taken its own good time. First, there was the breeding. We knew from our travels and analysis that Africa's demands on the new seed would be very different from those that worked wonders in Latin America or Asia. Second, there was no irrigation. Only a meagre 4 per cent of African farmland was irrigated, and it wasn't increasing with any great rapidity due to the very high cost of installation. We would simply have to focus on getting more out of rain-fed agriculture. As one farmer poignantly said, 'We don't have drought every year.' But even in the good rainfall years, the yields of farmers growing traditional crop varieties were still miserable. Third, Africa's sheer diversity of agro-ecologies was enough to boggle the mind. How does one breed for conditions as diverse as the Congo Basin, with its fishbowl-like humidity and depleted soils, the Sahel, with its suffocating heat interrupted by intermittent downpours, the highlands of East Africa, which throw at crops almost laboratory-specific conditions for the development of fungal diseases, and the coastal lowlands, with sandy soils and a propensity for droughts?

The problem was the crops themselves, or more precisely, the genetics within those crops, causing the sorghum varieties of the Sahel, for example, to grow 4 metres tall, and causing some local maize varieties to produce ears with a scant eight rows of grain. Many rice varieties also grew nearly 2 metres tall and their flowering was strictly controlled by daylength sensitivities again built into the plant's genetics. That genetic diversity was wonderful from a geneticist's perspective, and included within that diversity were some crucial traits for adaptation to local conditions, local taste and texture preferences, and diseases. But the overall crop itself was hopelessly out of synch with the need to feed rapidly increasing numbers of mouths. Few had ever sifted through that diversity to separate the valuable traits from the detrimental ones, and combine them with varieties of higher yield potential, which responded to fertiliser and other improved crop-management practices. The task appeared nothing short of monumental.

We decided to go local. A popular theme at the time among breeders was to involve farmers, themselves, in the design and then the selection of new varieties. We borrowed from that theme, but moved forward one generation and asked about the children of those farmers, who had probably spent a good portion of their childhood and adolescence planting and hoeing and harvesting the current generation of crops. What about the portion of that generation of farm-born and -bred young people who went to school and had studied agriculture? Wouldn't they have a pretty good notion of what constrained yields in their parents' fields? And if we could find them, and give them a top-notch education in crop genetics and breeding, wouldn't they then know both how to identify and how to deal with those constraints from a breeding perspective? And what if we could find enough such young people interested in pursuing such a career to cover all the major constraints of each major food crop? Might that not be the best way to deal with this seemingly endless diversity of crop breeding challenges?

After a few years we had a few good varieties. That is when we discovered we had a second, perhaps

It is one thing to have one bag of seed of a new crop variety ... It is quite another to deliver that seed to thousands and eventually millions of farmers living in isolated villages

even bigger, problem: We had no idea how to deliver the seed to farmers. It is one thing to have one bag of seed of a new crop variety that can transform yields and, with them, farmers' lives. It is quite another to deliver that seed to thousands and eventually millions of farmers living in isolated villages strung along dirt roads hundreds of kilometres from the nearest city. The Green Revolutions in Latin America and Asia happened during an era when governments planned and implemented huge national agricultural programmes. In Asia, the governments had largely delivered the new seed to farmers, along with fertiliser, through massive logistical efforts. But the world had changed since the 1960s. Government coffers around the world had been reduced to make way for the private sector. Putting our limited funds into public agencies at a time when they were being phased out made no sense. This Green Revolution, it seemed, would have to be driven by a private seed sector.

That meant building a seed industry from the ground up. Again, we decided to err on the side of local. We reasoned that, as the new varieties were bred and demonstrated to farmers and the public at large, local entrepreneurship would engage through the minds and actions of local business people, and we would have the opportunity to invest in the resulting seed businesses.

With few exceptions, that is how we now come to be working in partnership with over 80 public breeding teams working with 70 private, independent seed companies delivering approximately 40,000 metric tons of seed to farmers every year: local breeders plant demonstrations of new crop varieties to inform farmers of the new opportunities available through new seed, and local business people decide to risk their savings and assets on developing a business around production and delivery of that seed to local farmers. Annual sales of most of these companies are still below 1,000 metric tons but are rising steadily as we train their staff in production and processing techniques and train their leaders in seed company management.

It's a fascinating story of the local use of science to improve the lives of local people, most of whom are very poor, combined with local entrepreneurship for the benefit of the community. There is no telling yet whether this is a real revolution, but the results so far are very promising – most seed companies report selling out of their stocks every year and still being unable to meet demand for their seed. Meanwhile, the US Department of Agriculture (USDA) data show that crop yields in Africa have started to climb.

Stay tuned!

Dr Joe DeVries *is Director of the Program for Africa's Seed Systems (PASS), Alliance for a Green Revolution in Africa (AGRA), Nairobi, Kenya.*
Eden Square, Block 1, 5th Floor, PO Box 66773, Westlands, 00800 Nairobi, Kenya.
JDeVries@agra.org

Can the supply of quality seed match demand?

Dannie Romney, Roger Day, Daniel Karanja and Niels Louwaars

The Green Revolution in Asia was based on improved crop varieties (particularly rice and wheat) which, together with expanded use of agro-chemicals and irrigation, led to dramatic yield increases. But in Africa the revolution failed, with modest increases in production over the last few decades and for very few crops only. So will renewed efforts to get high-yielding varieties widely used in Africa be successful?

Developing new crop varieties, whether by traditional or modern breeding techniques or by genetic engineering, may be the easy part. We suggest that the challenge is to provide farmers with the opportunity to select varieties they want and need and that are most suited to their local circumstances and environment, so that their investment in quality seed is worthwhile. This requires both demand- and supply-side issues to be addressed.

What farmers want

Small-scale subsistence farmers are necessarily risk-averse, and in the circumstances in which many operate, high-yielding varieties (HYVs) are associated with higher risk. To achieve a good yield from an HYV a consistent supply of water is needed, but in Africa where most crops are rain-fed and rainfall can be erratic and unpredictable (and possibly more so as the climate changes) there has been little investment in irrigation. Reaping the benefits of purchased seeds also requires the use of other inputs such as fertilisers and agrochemicals, and these are often proportionately more expensive in Africa than in other regions. Farmers manage risk in different ways; for example, in western Kenya, farmers are more likely to plant hybrid maize in the long rains when there is lower risk of crop failure, but in the short rains, when risk of failure is higher, they use self-saved seed and plant more densely so that if the rains fail at least the crop can be used as fodder for their animals. In this context it is acknowledged that hybrid seed does not breed true in the next generation of saved seed, which does not deliver yield and performance in the way that the first generation does.

Even in a good season, if output markets function poorly, the returns on investment may discourage the use of HYVs. Selling the extra yield, beyond what is needed for consumption, can be constrained by lack of market information, poor infrastructure and price distortions.

Thus farmers look for various traits beyond high potential yield. The many different contexts in which farmers grow a crop mean that within one area different farmers may require different varieties at different times, so breeders and seed supply systems must be able to respond to these different demands.

Reaping the benefits of purchased seeds requires inputs such as fertilisers and agrochemicals

29

Seed supply systems

In the past, African governments took responsibility for provision of seed. Nowadays it is accepted that the private sector must be involved, but it commonly provides a very limited range of crop seeds. Multinational seed companies bring economies of scale, but may be less inclined than local companies to meet the heterogeneity of smallholder demand. Open-pollinated varieties are less profitable for companies, as farmers can save seed and avoid the recurrent cost of seed purchase. Commercial companies that invest in breeding tend to focus on species where they can make bigger margins by producing hybrids but where there is the recurrent cost of seed purchase. Local companies rely more on public-funded research for new varieties, but government research programmes may lack experience of forging the necessary partnerships to link public research with private opportunities to take varieties to farmers.

A variety of seed systems operate side-by-side in African countries. Local seed systems of saving and exchanging farm-produced seed provide by far the majority of cropped fields. Local commercial (often farmer cooperative) seed producers may produce locally adapted or modern varieties of a range of crops. National companies commonly depend on public varieties and hybrids. Multinational companies focus on hybrid maize and cotton, and for the most part commercial crops like tobacco or flowers, the market-partners providing the seed. Each farmer may use a variety of seed systems for different crops at any time.

Not only insufficient investment in plant breeding, but limited infrastructure and a lack of efficient and cost-effective marketing are the main constraints of small seed companies. So a recommended variety that is registered and in the market may still be unavailable to many farmers. The advice of extension agents must be matched to availability, because there is no point recommending seeds that farmers cannot afford to buy. Some methods that operate locally such as rural plant health clinics[1] or Farmer Field Schools may link to more widespread methods of sharing information such as radio and mobile phone, both widely used in African small-scale farming communities. For example, plant clinics operate in marketplaces and other locations readily accessible to poor farmers. Local extension staff, known as 'plant doctors', diagnose and advise on any crop with any problem. Information collected at clinics on the key problems farmers are facing can be used to develop key messages to be delivered at scale through complementary methods such as radio or mobile-phone-based systems.

A Green Revolution in Africa needs multiple demand-side and supply-side constraints to be addressed simultaneously

In most African countries seed production and marketing is regulated, normally by the national organisation responsible for plant protection. The aim is to ensure that only high-quality seed is sold; a farmer who invests in seed that does not germinate or grow as expected may lose money and is less likely to take the risk again. But over-rigorous regulation can stifle entrepreneurship and make provision of seeds that farmers want less profitable. For the different seed systems, different regulations should apply to support and control them. Self-regulation may be allowed for some varieties of seed, on the basis that a company selling bad seed will lose business and not survive. Full certification may be valuable when sufficient resources are allocated; a Quality Declared Seed (QDS) approach may be used to allow a wider group of stakeholders to participate in seed markets, but is not accepted in all countries.

Regional harmonization of seed regulation, such as in the Common Market for Eastern and Southern Africa (COMESA), should reduce the costs of bringing new varieties to market. But successful implementation depends on all countries having adequate regulatory capacity, and this may not yet be the case. In the area of genetically modified crops, some countries already have their biosafety frameworks and legislation established, but others do not. The patents that are associated with GM crops commonly put such varieties beyond the reach of smaller companies . . . and their customers.

Conclusion

A Green Revolution in Africa needs multiple demand-side and supply-side constraints to be addressed simultaneously. The Green Revolution in Asia was criticised for causing environmental damage of various types. This has led to calls for a 'Doubly Green' revolution or 'sustainable intensification', which will be even harder to achieve. But this is what the Comprehensive Africa Agriculture Programme envisages.[2]

Hard, but with the right policies and technologies, not impossible.

References
1   www.plantwise.org
2   www.nepad-caadp.net

Dr Dannie Romney is Global Director for Knowledge for Development at the Commonwealth Agricultural Bureau International (CABI), Nairobi, Kenya; **Dr Roger Day** *is Deputy Director for Development at CABI, Nairobi;* **Dr Daniel Karanja** *is a plant pathologist at CABI in Africa;* **Dr Niels Louwaars** *is Director of the Dutch seed association Plantum and a member of the Law and Governance Group at Wageningen University, The Netherlands.*
CABI Africa, PO Box 633, Nairobi, Kenya.
Wageningen Campus, Droevendaalsesteg 4, 6708 PB, Wageningen, Netherlands.

The Biosciences eastern and central Africa Hub

David White/dave@nospin.co.uk/ILRI

Segenet Kelemu

As a continent, Africa has vast natural resources, from precious metals and stones to plant, animal and microbial genetic diversity. Yet despite its natural wealth and invaluable contributions to the world's agriculture and global well-being, Africa still has 15 of the least developed nations in the world. According to the United Nations Educational, Scientific and Cultural Organization (UNESCO) *Science Report 2010*, Africa represents only 2.2 per cent of the world's researchers. Of this, sub-Saharan Africa (excluding South Africa) represents only 0.6 per cent. These

'The challenge of food security in sub-Saharan Africa is formidable, the timeframe for action is tight and the investment required is substantial. But the potential gains for human development are immense.'

Africa Human Development Report 2012: Towards a Food Secure Future, United Nations Development Programme

Current levels of agricultural R&D spending in sub-Saharan Africa are inadequate for agricultural growth ... and poverty reduction

figures are unacceptable if knowledge-intensive growth for more than a billion people is to be generated. There is no reason for Africa to continue to be a continent of glaring contradictions: a land of poverty amid plenty.

However, positive change is happening and there is so much to be optimistic about. In the past decade, a number of the world's fastest growing economies happen to be in Africa, and these have further expanded even at a time that some in Europe are struggling. This growth should be backed by a strong focus on and investment in people: a well-educated, skilled, motivated and well-nourished workforce for a knowledge-intensive and sustainable growth.

Investing in Africa's capacity to deliver its own solutions
Having studied and worked abroad for much of my career as a molecular plant pathologist, I returned to Africa in 2007 to help build African science leadership. I firmly believe that by fostering a knowledge-intensive growth approach to development, we can feed our continent and avoid future famine. In order to harness Africa's vast human capital, there is a need to invest in people and to ensure that more African scientists devote their talents to addressing regional challenges.

Not only do we need to foster the talent that lies within Africa, we need to retain it. One way of doing this is to create an enabling environment for cutting-edge science through substantial and sustained investment in science, technology and innovation. Current levels of agricultural research and development (R&D) spending in sub-Saharan Africa are inadequate for agricultural growth, establishment of food and nutritional security and poverty reduction.

Studies show that investments in agricultural R&D generally provide high returns. Average rates of return have been documented in the range of 35 per cent for sub-Saharan Africa to 50 per cent in Asia in several studies. Without increased investment, Africa will continue to see ground-breaking discoveries in crops built on plant traits from African natural resources being conducted outside the continent, largely due to the shortage of bioscience researchers and adequate infrastructure on the continent.

Despite these challenges, the exciting journey to build Africa's science capacity has begun. A wide range of programmes is under way to build and strengthen Africa's agricultural research for development establishment and human capacity. In collaboration with the international R&D communities, we are seeing some real progress and exciting advancements for Africa.

The *Biosciences eastern and central Africa – International Livestock Research Institute Hub* (BecA–ILRI Hub) is one of many such initiatives. It is a leading biosciences research and capacity-building centre for Africa. Based in Nairobi, Kenya, its mission is to improve the livelihoods of resource-poor people in Africa through the development and use of new biosciences technologies. All activities focus on developing and producing technologies that help poor farmers to improve their productivity and income, secure their assets and increase their market opportunities.

The BecA–ILRI Hub was developed within the framework of the New Partnership for Africa's Development (NEPAD)/African Union African Biosciences Initiative. The establishment of the BecA network, alongside three others – the North Africa Biosciences Network, the Southern Africa Network for Biosciences and the West Africa Biosciences Network – is in line with the aim of the Comprehensive Africa Agriculture Development Program (CAADP): to help

African countries reach a higher path of economic growth through agriculture-led development.

The BecA–ILRI Hub serves 18 countries of Eastern and Central Africa, namely Burundi, Cameroon, Central African Republic, Congo Brazzaville, Democratic Republic of the Congo, Equatorial Guinea, Eritrea, Ethiopia, Gabon, Kenya, Madagascar, Rwanda, São Tome and Principe, Somalia, South Sudan, Sudan, Tanzania and Uganda. While keeping a focus on its mandate in the region it is responding to the growing demand for its services across the whole of Africa.

Specifically, the BecA–ILRI Hub provides a common biosciences research platform, research-related services and capacity-building and training opportunities. These activities provide a focal point for the African scientific community and the aim is to support the activities of national, regional and international agencies and staff as they address agricultural problems of the highest priority for alleviating poverty and promoting development.

World class research facilities

In May 2003, Canada's commitment of CAD$30 million from the Canada Fund for Africa enabled the establishment of the BecA laboratory facilities; bringing Africa's research capability up to par with that of the world's most developed countries. Africa's scientists can now comfortably venture into new realms of science, using cutting-edge equipment without the constraints of inadequate laboratories, and without having to incur prohibitive costs and face restrictive regulations to conduct the same research overseas.

Many scientists have already used the capacities of the BecA–ILRI Hub in the crop, livestock and microbial sciences. The biosafety level containment laboratory, BSL-3, is one of a handful in Africa for research on animal pathogens constraining livestock health in Africa. In order to serve the region's full agricultural improvement needs, the BecA–ILRI Hub also has state-of-the-art crop and microbial research laboratories. These include a non-containment tissue culture laboratory, BSL-2 plant transformation laboratories and a comprehensive bioinformatics platform for genome and meta-genome sequencing, as well as a greenhouse complex including three BSL-2 greenhouse compartments.

Building African science leaders

In order to address the lack of sufficient expertise in science and technology in Africa, a strong programme in capacity-building and training is central to the success of the BecA–ILRI Hub initiative. A highly skilled, healthy and well-paid workforce is critical in making Africa productive and globally competitive.

The Africa Biosciences Challenge Fund (ABCF) is a new and innovative way of enhancing African biosciences capacity while tackling agricultural constraints. It was established as part of the BecA–Commonwealth Scientific and Industrial Research Organisation (CSIRO) partnership with initial funding from the Australian Agency for International Development (AusAID). Subsequently, the Bill & Melinda Gates Foundation and the Swedish Ministry for Foreign Affairs through the Swedish International Development Cooperation Agency (Sida) have become valued contributors.

The Syngenta Foundation for Sustainable Agriculture, an early technical and financial supporter of the Hub, also contributes core support to the ABCF initiative and other activities that help make the facilities accessible to the greater

In order to address the lack of sufficient expertise in science and technology in Africa, a strong programme in capacity-building and training is central to success

37

African scientific community. Research and capacity-building under ABCF is conducted via hands-on training workshops and provision of research fellowships to early-career African scientists with placements for periods of three to six months at the Hub to undertake research projects. The ABCF also funds institutional capacity-building and visits to BecA countries to raise awareness about the Hub, helping identify talents and agricultural research constraints.

Between 2007 and 2011, the BecA–ILRI Hub facilities and skills have been accessed by 316 graduate students and visiting scientists. Furthermore, 1,733 young Africans have received training in various biosciences-related fields at 54 events during this period. This demand is expected to grow significantly in the coming years as the Hub diversifies its own services and capabilities and strengthens relationships with international and African R&D institutes. The Hub has been identified as one strong networking model for making progress in capacity-building coupled with research that can lead to enhancing the productivity of the agricultural sector.[1]

Most importantly, the BecA–ILRI Hub enables collaborative research amongst African and international scientists focusing on neglected and under-researched areas that address food security. Agricultural research systems in sub-Saharan Africa are fragmented into nearly 400 distinct research agencies across 48 countries. Centres like the Hub facilitate synergies of skills and knowledge and create efficiencies in R&D in parts of Africa, thus ensuring that maximum benefits are realised for farmers and communities.

A highly skilled, healthy and well-paid workforce is critical in making Africa productive and globally competitive

It is my strong belief that the BecA–ILRI Hub model provides a platform for success that can be duplicated in other regions across Africa. By advancing the science and technology capacity through institutions, we can help to provide the science that addresses food and nutritional insecurity. Initiatives like BecA are creating the next generation of technological innovators who will lead the coming agricultural revolution from within Africa, enabling a bright future for the continent.[2]

Global support from investors
BecA-ILRI Hub's global investors include AusAID, the Bill & Melinda Gates Foundation, the Syngenta Foundation for Sustainable Agriculture, the Swedish Ministry for Foreign Affairs (through Sida) and the Canadian International Development Agency.

References
1   www.asti.cgiar.org/pdf/conference/Theme2/Moock.pdf
2   http://hub.africabiosciences.org

Dr Segenet Kelemu is Director of the Biosciences eastern and central Africa (BecA) Hub, Nairobi, Kenya.
BecA Hub, International Livestock Research Institute (ILRI), PO Box 30709, Kabete Campus, Old Naivasha Road, Nairobi 00100, Kenya. SKelemu@agra.org

Training for the future of food security

S. Noorani/Majority World/Still Pictures

Eric Yirenkyi Danquah

Today, about 1 billion people globally lack access to adequate food. Food insecurity and economic crises have drawn attention to the urgent need and potential for developing sustainable agricultural systems. By 2050, the global population will reach 9 billion and the demand for food is expected to double.[1] At the same time, the world's agricultural systems will be increasingly challenged by changing climates. In sub-Saharan Africa, dramatic advances are required in food production (up to 100 per cent increases and possibly more in yields of food crops) to address food needs.[2] This will require the integration of conventional plant breeding and biotechnology approaches to develop new crop varieties for sub-Saharan Africa, where yields of food crops are the lowest in the world and increased productivity is a necessary condition for food security.[3]

West and Central Africa (WCA) together with the Horn of Africa are perhaps the most food-insecure regions in sub-Saharan Africa. Agriculture has long been identified as the engine for economic growth and development in this region but the impact to date has been far from satisfactory. Unlike the successes of the Green Revolution, achieving widespread impact with a few new crop varieties has a very low probability of success in WCA as the context is so different from that of Asia and Latin America. In WCA, 95 per cent of the food is grown as rain-fed agriculture. Crop production is therefore subjected to the vagaries of the weather together with biotic (e.g. pests and diseases) and abiotic (e.g. radiation, temperature and water) stress challenges. The diverse agro-ecologies make crop breeding challenging in an African context. This is complicated by the structure of African farming systems: the diversity within and between countries demands a variety of locally adapted staple crops often cultivated by predominantly small-scale farmers.

WCA faces a major challenge in feeding close to half a billion people. Furthermore, the forecast from the United Nations is that the population there will more than triple in the next decade, and WCA will become one of the most vulnerable places in the world unless action is taken now. Ensuring food security in WCA calls for concerted regional and international effort to increase farm productivity. This requires the political will to provide improved rural infrastructure, access to more effective inputs, storage of farm produce to minimise post-harvest losses, transportation, processing and access to markets. Effecting such transformational changes will entail local entrepreneurship to drive innovation across the agribusiness sector.

... the need for a critical mass of scientists trained in plant breeding with conventional and molecular expertise is urgent

The development of new and improved crop varieties with in-built resilience to climate change coupled with resistance to pests and diseases

41

is required. Given the crucial role that genetic improvement through plant breeding can play to provide vital inputs in boosting crop productivity, the need for a critical mass of scientists trained in plant breeding with conventional and molecular expertise to develop superior varieties is urgent. The dearth of plant breeders with such skills in WCA is well documented. Given that the majority of practising plant breeders will retire in the next decade, there is an urgent need to train more plant breeders. A way of estimating the current needs for plant breeders is to assume that there should be at least one trained breeder for each crop in every agro-ecological zone in each country in WCA.

The promise: WACCI – a network of locally trained plant breeders

The West Africa Centre for Crop Improvement (WACCI) was established in June 2007 at the University of Ghana to train plant breeders for Burkina Faso, Cameroon, Ghana, Mali, Niger and Nigeria. For the six countries approximately 324 plant breeders are needed for the nine major crops. It is estimated that about 50 breeders are active in these countries. WACCI has 54 students doing their PhDs in plant breeding who will enter the profession in the next six years, leaving a deficit of some 220 breeders. One way forward is to establish partnerships, exchanges and collaborations with major institutions and departments in the more developed countries.

At WACCI, the students undertake thesis research on staple crops in home institutions after two years of coursework at the university. The students engage farmers in discussions in a pre-research phase and incorporate farmers' challenges into their research. The core concept of the programme is farmer-driven with students embedded in the key

The financial obligations are enormous but the potential benefits to present and future generations far outweigh the investments

agro-ecological zones in their home countries for their research. The expected outputs are qualified, competent and motivated plant breeders applying the full scope of breeding methodologies to address local crop constraints. The result will be enhanced food security through the delivery of superior crop varieties that meet the needs of farmers and a high retention rate of breeders in Africa.

The WACCI concept has been recognised by the West and Central African Council for Agricultural Research and Development (CORAF/WECARD) and the Chicago Council on Global Affairs. It has been adopted by the Agricultural Research Council of Nigeria to be implemented at the Centre for Crop and Animal Improvement to be located at the Ahmadu Bello University in Nigeria.

The training of a new generation of plant breeders at WACCI will build human capacity and in the long term contribute to food security in sub-Saharan Africa. The urgent need to sustain WACCI as a Regional Centre of Innovation to train plant breeders beyond 2017 is more pressing than ever. The financial obligations are enormous but the potential benefits to present and future generations far outweigh the investments.

The African Centre for Crop Improvement (ACCI), founded on a similar concept at the University of KwaZulu-Natal five years before WACCI, has graduated plant breeders who continued their research after graduating and released varieties which have been adopted by farmers in Eastern Africa. Graduates have already released 23 new improved clones, cultivars and hybrids in Uganda, Kenya, Malawi, Zambia, Mozambique and Burkina Faso. For example, annual yields of cassava have been doubled while reducing the time to maturity from 18 months to 7 months and dry bean varieties totally resistant to weevils have been released. The Alliance for a Green Revolution in Africa

43

(AGRA) is commended for the audacity to fully fund PhD programmes in Africa. WACCI has guaranteed funding until 2017.

An endowment fund is urgently needed to sustain WACCI and hence train plant breeders required to develop superior genetic varieties for West and Central Africa.

References

1   FAO (2010) The State of Food Insecurity in the World: Addressing Food Insecurity in Protracted Crises. Rome: Food and Agriculture Organization of the United Nations and World Food Programme.
2   The Montpellier Panel (2012) Growth with Resilience: Opportunities in African Agriculture. London: Agriculture for Impact.
3   Conway G., Waage J. (2010) Science and Innovation for Development. London: UK Collaborative on Development Sciences.

Professor Eric Yirenkyi Danquah is a Professor of Plant Molecular Genetics at the Department of Crop Science, University of Ghana.
West Africa Centre for Crop Improvement, PMB 30, University of Ghana, Legon, Ghana.
edanquah@wacci.edu.gh

Biofortified sorghum: lessons for biotechnology

D.Harms/Wildlife/Still Pics/Specialist Stock

Florence Wambugu

The Africa Biofortified Sorghum (ABS) Project started as part of the Bill & Melinda Gates Foundation's Grand Challenges in Global Health (GCGH). The project remains true to the Grand Challenges vision of using a 'specific scientific or technological innovation to remove a critical barrier or solving an important health problem in the developing world that has a high likelihood of global impact and feasibility'.

The project seeks to develop a more nutritious and easily digestible sorghum variety. Deficiency in vitamin A is one of the most prevalent problems in sub-Saharan Africa. Severe vitamin A deficiency has very high mortality rates (60 per cent), but even sub-clinical deficiency is associated with a 23 per cent increase in pre-school-age mortality in areas with endemic vitamin A deficiency.[1] It is this urgency and likely impact that has informed the project's new focus on vitamin A.

Initially, multiple interventions were envisaged. This included increasing levels of essential amino acids (especially lysine) and micronutrients such as iron and zinc and vitamin A. Following the successful development of the 'Golden Sorghum' fortified with pro-vitamin A (beta-carotene), all other interventions were put on hold; the focus is now to develop an early, or first product, with increased levels of pro-vitamin A.

After the initial five-year funding, Du Pont Pioneer secured additional funding from the Howard G. Buffett Foundation (HGBF) to focus on critical technology milestones while Africa Harvest leveraged internal funds to continue with critical non-technical milestones. In less than seven years since the project was initiated, nearly a dozen field trials have been done. Confined field trials (CFTs) have been successfully carried out in the USA, Kenya and Nigeria.

As one of the four agricultural projects funded by the Bill & Melinda Gates Foundation the success and challenges of the project provide important lessons for the future.

Developing African scientific capacity and leadership

The project has contributed significantly to the strengthening of African scientific networks. It pioneered the innovative model of a mutually beneficial relationship with Western scientific institutions. This was achieved through a multidisciplinary core team that helped integrate 13 African and international institutions that formed the ABS consortium. A compelling vision helped African institutions overcome the perception that they would play the role of junior partners. South Africa's Council for

A compelling vision helped African institutions overcome the perception that they would play the role of junior partners

Scientific and Industrial Research (CSIR) took R&D leadership. It was the institution through which technology was channelled to the consortium and the continent. Other institutions were identified based

The project played a major role in assisting in the development of strong and effective regulatory policies in several countries

on extensive evaluation of infrastructural and scientific capacity. DuPont Pioneer was particularly attracted by the idea of working with an 'Africa-led' consortium and remains committed to the success of the project. Looking back, enormous individual and infrastructural capacity was built, especially with the support of DuPont Pioneer, who hosted many African scientists at their world-class facilities in Johnston, Iowa.

The critical role of the private sector

It was through my membership on the DuPont Biotech Advisory Panel that I learnt of their successful use of corn gene delivery protocols for sorghum. They had tested the expression of a gene of interest (high lysine storage protein) and several transgenic lines had been created. Three of them had up to 50 per cent increases in lysine. It was this early work that formed the basis of the ABS Project proposal. Later, as the project was implemented, it was DuPont Pioneer's private-sector culture and discipline that enabled the project to make great and speedy progress. Their experience was critical in many ways; perhaps the most important was their ability to shift efforts from a 'science-for-science's-sake' approach to focus the team on producing a viable product.

Intellectual property

As a consortium member, African Agricultural Technology Foundation (AATF) undertook early intellectual property (IP) audits to ensure that no compliance issues were overlooked and that the project had Freedom to Operate (FTO).

Donated IP was checked against the Organisation Africaine de la Propriété Intellectuelle (OAPI) and the African Regional Intellectual Property Organisation (ARIPO). To date, all project IP remains in the public domain and Africa Harvest holds it in trust as a public good.

Biosafety and regulatory capacity

When the project started, only South Africa, Burkina Faso and Egypt had commercialised GM products. The project however played a major role in assisting in the development of strong and effective regulatory policies in several countries. More specifically, the ABS Project's role in the passage of the biosafety laws in Kenya and Nigeria are well documented.

Communication and issues management

The ABS Project pioneered the development of Africa-specific communication tools and stakeholders engagement. Early in the project, Africa Harvest carried out a survey of public perceptions of GM crops to help inform the project's communication strategy and also design how to address biosafety and regulatory issues. Overall, public acceptance of GM technology has increased in Africa. For example, the New Partnership for Africa's Development (NEPAD) has provided leadership through its position on Africa's need for Freedom to Operate. At project level, many biotech projects have built communication protocols based on the ABS Project's experiences.

Balancing commercial interests and country needs

According to Bill Gates, 'it's shocking how little research is directed towards the diseases of the world's poorest countries'. In Africa, it is widely acknowledged that the continent's lack of enthusiasm to embrace GM technology was partly rooted in the lack of research targeting local crops. The ABS Project plays an important role in getting political buy-in, especially in countries where sorghum plays a critical role in mitigating challenges related to malnutrition and climate change.

In the early days, the consortium's biggest challenge was framing malnutrition as a 'disease of the poor', just as obesity has been described as a disease of the rich. Africa's health challenges are viewed through the lenses of major diseases such as malaria, tuberculosis and HIV-AIDS. The project has successfully managed to link nutrition with the continent's burden of disease. For example, successful HIV-AIDS interventions, such as anti-retroviral, work best when those taking the drugs have good nutrition.

Project and consortium management

Given the large number of institutions involved in the project, Africa Harvest had the role of ensuring that contractual issues were dealt with effectively, that every organisation understood and delivered on its milestone-based strategy, and that a single consortium culture glued things together. An External Advisory Board helped the project get fresh perspectives on how best to implement the project. Among the scientific issues the Board helped identify and deal with were the extent to which gene flow occurred and the need to identify at an early stage locally adapted and high-yielding sorghum varieties.

Funding

The project was very well funded during the first, five-year phase which started in 2005. One lesson learnt after non-extension of funding for phase two is that support from multiple sources is more sustainable than relying on a single funder. Despite funding challenges, in the last two years technological advances have been

... support from multiple sources is more sustainable than relying on a single funder

49

made and there has been a paradigm shift in how Africa views GM crops. The enabling policy framework, good governance, infrastructure and scientific capacities have improved. Societal concerns are being addressed and creative funding structures are being explored. It is anticipated that the nutritionally improved sorghum will be available to farmers in the not-too-distant future.

References
1   McGuire J. (1993) Addressing micronutrient malnutrition. *SCN News* No. 9. Geneva, Switzerland: United Nations Subcommittee on Nutrition (ACC/SCN).

Dr Florence Wambugu is the CEO and Founder of Africa Harvest Biotech Foundation International (Africa Harvest, www.africaharvest.org).
3rd Floor, Whitefield Place, School Lane, Westlands, 00621 Nairobi, Kenya.
fwambugu@africaharvest.org

Achieving water efficiency with maize

Denis T. Kyetere, Sylvester O. Oikeh and Grace Wachoro

Increasing incidences of drought in Africa are affecting maize production, one of the major staple crops for a majority of the population. This has resulted in frequent crop failures leading to hunger and poverty in sub-Saharan Africa. One of the most challenging problems in plant breeding today is that of improving drought tolerance. A public–private partnership known as the Water Efficient Maize for Africa (WEMA) consortium was formed in 2008 and has been working to develop drought-tolerant maize for smallholder farmers in sub-Saharan Africa with promising results that indicate that maize can be made more water efficient by modern methods of plant breeding.

Effects of drought on maize production

Incidences of drought in the world are increasing. This is especially so in Africa, which is now referred to as a drought-prone continent where the vast majority of agriculture is rain-fed. Drought leads to crop failure, hunger and poverty, and climate change is worsening the problem. Recurring droughts – more than ten drought events between 1970 and 2004 – are a persistent challenge in sub-Saharan Africa, making farming risky for millions of smallholder farmers and their families who rely on annual rainfall to grow their crops. Drought stress is one of the top two factors responsible for limiting maize production[1] and it is estimated that in sub-Saharan Africa it causes yield losses of 10–25 per cent on average.

Maize is the most widely grown staple crop in Africa and it provides food for more than 300 million people. Some countries in Eastern and Southern Africa rely on maize for more than half of their total calorie consumption.[2] However, the average maize yield per hectare for a farmer in sub-Saharan Africa is much less than that of farmers in the USA and Europe. Smallholder farmers in Africa are unable to utilise basic technologies like improved seed and fertiliser due to lack of capital or because they are unwilling to invest the little capital they have for fear of losing their investment during drought.

The Water Efficient Maize for Africa project

Due to both the complexities of drought itself and the plant's response to abiotic stresses such as moisture depletion, a single organisation is unlikely to be in a position to address all the challenges of developing drought-tolerant maize, or that a single technology or methodology will provide a solution. The Water Efficient Maize for Africa (WEMA) partnership is taking multiple approaches and utilising resources across organisations to reach the project's objectives. The focus is threefold: to develop new varieties (germplasm) using conventional breeding techniques, e.g. production of doubled haploid plants or marker-assisted recurrent selection (MARS); to undertake discovery breeding to identify genes that confer drought tolerance in the maize

Both conventional and molecular breeding programmes have resulted in positive developments

genome; and to test the introduction of drought-tolerance transgenes (GM) into adapted varieties in the partner countries.[3] The aim is to produce WEMA products that will be drought-tolerant (white single-cross and/or three-way cross hybrids), giving at least a 20–35 per cent yield advantage under moderate drought conditions compared to commercial hybrids produced in 2008.

WEMA is a public–private partnership whose objective is to develop drought-tolerant maize and make it available royalty-free for the benefit of smallholder farmers in sub-Saharan Africa. The project, which is being implemented in five countries of Eastern and Southern Africa, aims to improve food security and rural livelihoods among smallholders by developing hybrid maize varieties that can tolerate conditions of moderate drought.[3] Currently WEMA is field-testing Monsanto's drought-tolerant lead variety (MON87460) which has an introduced transgene (*CspB*) and an inducible cold shock tolerant gene from *Bacillus subtilis* which confers improved adaptation to stress in several plant species. The transgene is expected to provide additional gains in drought tolerance.

The partnership arrangements of the consortium are shown in Box 1 and financial support has come from the Bill & Melinda Gates Foundation and Howard G. Buffett Foundation (2008–2012).

Progress to date

Both conventional and molecular breeding programmes have resulted in positive developments in the last four years. Under conventional breeding, 40 drought-tolerant varieties (not GM) were submitted to the National Performance Trials in Kenya (16 varieties) and Uganda (24 varieties) in 2012. These trials are used to determine the suitability of new varieties to a country's growing conditions. Tanzania and Mozambique are set to do the same soon. Kenya and Uganda are currently carrying out the third round of transgenic (GM) confined field trials while

Box 1. Contribution of partners to the WEMA project

Organisation	Expertise	Contribution
African Agricultural Technology Foundation (AATF)	• Public–private partnership management • Legal affairs and intellectual property management • Seed system operations • Stewardship of conventional products in smallholder settings • Communications • Regulatory affairs • Project leadership	• Partnership management • Licensing, intellectual property and legal affairs management, including facilitating negotiations of all agreements • Product deployment and stewardship strategy • Project communications, outreach and awareness • Regulatory approvals and compliance • Project monitoring and evaluation
International Maize and Wheat Improvement Center (CIMMYT)	• Conventional, abiotic and biotic stress breeding • Conventional, molecular and doubled haploid breeding	• Sub-Saharan Africa adapted drought-tolerant germplasm • DNA marker information
Monsanto	• Conventional, molecular and doubled haploid breeding • Biotechnology testing and stewardship • Seed production, deployment and licensing	• Global germplasm • DNA marker information • Transgenic traits for drought tolerance (with BASF) and insect-pest protection • Seed deployment strategy
National Agricultural Research Systems (Kenya, Uganda, Tanzania, South Africa and Mozambique)	• Conventional breeding • Field-testing and regional trials • Knowledge of farmers' product needs	• Locally adapted germplasm

South Africa is in the fourth trial. Mozambique and Tanzania continue to work towards securing regulatory approvals to conduct transgenic trials.

During the first four years of research, insect infestation, especially maize stem borers, was identified as a major potential threat to the anticipated benefits of drought-tolerant maize varieties. To overcome this challenge, the project is adding insect-pest protection with the introduction of genes (*Bt*, *Cry1Ab*) to WEMA-developed varieties to enable farmers to secure an insect-pest-resistant maize crop through healthier plants that are able to use the water and nutrients more efficiently during drought stress.

The first WEMA varieties developed through conventional breeding could be available by 2014. Varieties developed using transgenic approaches will be available to farmers depending on research results and regulatory approvals in each of the WEMA countries. Farmers could have access to these drought-tolerant and insect-pest-protected maize varieties within six to seven years.

Regulatory compliance

In order to conduct confined field trials and eventually commercialise the maize varieties, there must be enabling biosafety laws in place. Only two project countries – Kenya and South Africa – have functional biosafety laws that allow for testing and commercialisation of biotech crops. The project depends on policy-makers in the various countries to put in place the necessary regulatory frameworks so that the farmers of those countries do not miss out on the benefits that biotechnology can bring to enhancing agricultural productivity.

Creation of awareness

Meeting the technical milestones of the WEMA project also requires buy-in, ownership and support of key stakeholders. However, the introduction of biotechnology crops into Africa is controversial, compounded by social, ethical and political considerations. The WEMA communication and outreach strategy has sought to ensure a good understanding and appreciation of the project, its partners and the technology that will support the safe transfer, adaptation and delivery of the maize varieties. The project supports the creation of awareness on biotechnology among all stakeholders for informed decision-making.

Lessons learned

In the course of the four years that the project has been in operation, some lessons learned have informed implementation of activities and will be valuable as the project progresses into the next phase. The involvement of the National Agricultural Research Systems (NARS) in breeding promotes opportunities and builds a foundation in crop improvement in the partner countries. The approval of applications to conduct confined field trials in some of the countries has enabled the project to make progress towards delivering the much-needed drought-tolerant maize varieties to smallholder farmers. This can be attributed to the capacity-building support that the project has given regulators in risk assessment and decision-making.

Way forward: deployment of the maize varieties

To ensure that the drought-tolerant and insect-pest-protected maize varieties from the WEMA project are available to smallholder farmers in sub-Saharan Africa, strategies are already being put in place for the deployment phase. This phase requires financial support from funding agencies so as to strengthen the local seed industry to produce quality drought-tolerant varieties and promote the delivery of the products to smallholder farmers. Most local seed companies currently lack the expertise and facilities necessary to produce certified seed of the quality and volume anticipated for farmers to realise the benefits of the drought-tolerant and insect-pest-protection traits. A further problem is that hybrid maize seed cannot be saved from one year to the next and will have to be purchased by or donated to farmers every year.

Conclusion

The project has made good progress since its launch and is on course to deliver drought-tolerant and insect-pest-protected locally adapted maize varieties to smallholder farmers in sub-Saharan Africa. This has been largely possible because of the excellent collaboration among all the partners. The regulatory environment in South Africa, Kenya and Uganda has enabled confined field trials to be conducted in these countries and the results from the trials are promising. The project is looking forward to moving on to the next deployment phase that will see the varieties delivered to smallholder farmers.

References

1   Heisey P.W., Edmeades G.O. (1999). Maize production in drought-stressed environments: technical options and research resource allocation. In *World Maize Facts and Trends 1997/98*. Mexico, D.F.: CIMMYT

2   Banziger M., Diallo A.O. (2001). Stress-tolerant maize for farmers in sub-Saharan Africa. In: *The Maize Program: Maize Research Highlights 1999–2000*. Mexico, D.F.: CIMMYT.

3   Oikeh S., Ngonyamo-Majee D., Mugo S., Mashingaidze K., Cook V. (2012). WEMA project as an example of a private–public partnership. *Biotechnology in Agriculture and Forestry* **67** (in press).

Dr Denis Kyetere is the Executive Director of the African Agricultural Technology Foundation (AATF); **Dr Sylvester O. Oikeh** *is the Project Manager of the Water Efficient Maize for Africa (WEMA) project;* **Grace Wachoro** *is the Corporate Communications Officer (AATF).*
African Agricultural Technology Foundation (AATF), PO Box 30709, 00100 Nairobi, Kenya.
D.Kyetere@aatf-africa.org; S.Oikeh@aatf-africa.org; G.Wachoro@aatf-africa.org

Where will the water come from?

KickStart International

Nick Moon

Africa has millions of acres of underutilised arable land and millions of smallholder farmers ready to work on it, if it is really worth their while. So why are we still unable to fully exploit this opportunity? How come most smallholder families produce so little and remain net importers of food into their households?

Many solutions have been developed and demonstrated for the smallholder, but we will not accomplish the 'African Green Revolution' unless we overcome our unsustainable dependence on rain-fed agriculture. A staggering 97 per cent of smallholder agriculture is rain-fed, an increasingly risky strategy in these times of accelerating climate change. Most parts of Africa have only one annual rainy season, although some get two. So food production is limited from seed to harvest in only four to five months, or at best eight or nine months out of 12, a huge waste of time and resources. Furthermore, when everyone harvests and markets at the same time, farmers get rock-bottom prices. A few months later when they need to buy food, they must pay top dollar. So they spend their money to survive dry periods, and have nothing remaining to invest in inputs come the next rainy season. Tragically, some 40 per cent of produce is lost post-harvest because existing infrastructure is overloaded at peak times of the year, only to be underloaded at others. The answer here is clearly irrigation.

To synchronise production more closely with demand and achieve a smoother, more efficient 'just-in-time' supply chain calls for big investments in responsible agricultural water management (AWM). Contrary to popular opinion, much of Africa is not water-challenged, although water resources are not evenly distributed around the continent. Some 97 per cent of the annual rain that falls on Africa flows unchecked into the ocean. If we captured and managed all of it there would be enough water for everyone on the planet! Meanwhile, recent research has shown that truly massive reserves of water lie in African underground aquifers. So we have plenty of water, as well as land and farmers. How do we bring these components together so as to maximise the value we can create with new seeds derived from high-quality genetics?

KickStart[1] (a social enterprise dedicated to designing and promoting productivity-enhancing technologies for the very poor at the 'base of the pyramid') has developed both an AWM technology for smallholders, and an effective 'last mile delivery' strategy. The standard policy is to invest in highly capital-intensive water catchment and irrigation schemes. A serious drawback is the astronomical expense: both initial capital (often US$10,000 per irrigated hectare) and the continued reliance for operations on costly specialised management and

... we must develop water management to exploit the value of new seeds from modern genetics

technical support. Another is that by their nature these schemes are concentrated on specific site developments that have disruptive, often negative, impacts on local environments and cultures. Additionally, because these projects commonly use channel/furrow or flood irrigation systems, water consumption per hectare or per crop is wastefully high due to evaporation and plant transpiration together with poor water retention properties of the soil.

Irrigation need not be like this. A number of very low-cost AWM technologies, specially designed to fit the socio-economic and cultural circumstances of smallholder farmers, and highly efficient in terms of energy and water consumption, have been developed and proven. More crop-per-drop micro-drip systems are a well known example. KickStart's low-cost, human-powered machines enable very poor people to use their time and energy more productively, add value to their existing skills and assets, and make more money. 'MoneyMaker', a pressure-hose irrigation pump, costs the farmer US$100 for the larger treadle-operated model, or US$40 for the smaller manual 'hip pump'. They can pump water up from rivers and streams, lakes, ponds, dams, small water catchments and hand-dug wells from depths of 8 metres/25 feet. They can then pressurise the water upwards, for a total suction and pressure head of 14 metres/45 feet – the height of a four-storey building. Adding the cost of inlet pipes and outlet hoses, a farmer is asked to make a capital investment ranging from US$60 to US$150. Running costs are US$7 annually for replacement parts, and no specialist expertise is required. The pump system is managed by two people (usually husband and wife). One operates the pump (by treadle or handle) while the other holds the far end of the delivery hose and

Farmers are concerned about how to raise their families, send the children to school, obtain healthcare, and save up to buy a dairy cow ...

directs the pressurised water straight onto the root system or onto the plant itself. This form of plant watering enables farmers to use far less time and energy to deliver a lot more water to a lot more plants, saving hours of drudgery drawing and distributing water by bucket and rope, or waiting and praying for rain. Altogether between 10 and 18 hours per week of pump operation suffices to adequately irrigate up to 0.75 of a hectare, depending on pump type and other conditions. Close to 20 million rural families in Africa live in places where these pumps can be installed and 200,000 smallholders use them already, having bought one from a local store, or obtained one through a development programme. Impact evaluations show that MoneyMakers empower smallholders to transform subsistence farms into commercial enterprises, growing more food each year, and switching to higher nutritional value crops that command better prices in the market. Put simply, a US$100 investment typically generates additional net household incomes of US$500–1,000 per year over the five-to-nine-year life of the pump. The cost of this solution per irrigated acre compares very favourably with large schemes. The farmer invests just US$100 to put an acre (or US$250 per hectare) under effective water management, versus US$5,000–10,000 for large schemes.

Why aren't 20 million African farmers already using this sustainable system? There are three basic reasons: they don't know about it; they have not yet seen the value it offers; or they cannot easily get a pump for lack of capital to invest or want of a nearby stockist.

None of these constraints need be there. Already, with very modest resources, KickStart has organised the mass production, shipping and distribution (through a wholesale and retail network of 500 outlets in 15 countries), the marketing, promotion and ultimate sale of 200,000 MoneyMaker pumps, which are now being used to water-manage 80,000 hectares. All this has been achieved with very little help from government ministries and extension services, or international agencies active in smallholder agriculture, who say they baulk at apparently endorsing a proprietary brand of pump. Their interventions are often delivered beside, not through, the market – 'demonstration projects' are common and mostly unsustainable.

Most people see African smallholders as victims of economic forces beyond their control who need hand-outs. And yet they do survive; 130 million families (650 million people) live not just without subsidy but are heavily taxed as well. However, it is true that selling new machines to subsistence farmers has not been easy and sales in the first year were only 300 pumps (last year 30,000). Smallholders are very difficult to reach in their deep rural areas where there are no paved roads or electricity. They are deeply conservative, very cautious and highly risk-averse. The value proposition we take to them is personal and familiar: buy this, use it, make money and provide for your family. Poor farmers deep in rural Africa are not anguishing over issues of global food security. Nor will they be very responsive to patriotic exhortations to exercise their civic duty and feed the nation – they are too poor and hungry for that. Farmers are concerned about how to raise their families, send the children to school, obtain healthcare, and save up to buy a dairy cow, a solar panel, or pay for a daughter's college education. And like everyone else, they want to enhance their social status, be someone in the eyes of the community.

KickStart's marketing and sales strategy focuses on these priorities. We show and tell, demonstrate MoneyMaker pumps at fairs and markets and at farmers' field days with testimonies and refer-

... we have lowered the barriers to entry for the very poorest farmers

rals from satisfied users, and promote them on local-language FM radio channels. All this gradually overcomes the initial doubts and suspicions of farmers. Also, by unconditionally guaranteeing the pump, by co-marketing it with seed and fertiliser and crop protection products, and by making it possible for farmers either to obtain loans from local banks or to pay gradually through a series of mobile phone money transfers, now common in rural populations, we have lowered the barriers to entry for the very poorest farmers to get involved.

So what does this teach us? If we are to ensure food and nutrition security in Africa (and grow economically) we must develop and market AWM (catchment and irrigation) technologies which are economically, environmentally and culturally appropriate for use by smallholder farmers, and which enable them to use this precious – but infinitely renewable – resource effectively to exploit the value of newly developed seeds and modern genetics. And were governments and development agencies to recognise and support smallholders in the same way as other prospective business investors, that would help too.

Reference
1   Kickstart at: http://www.kickstart.org

Nick Moon was Co-Founder and Managing Director, KickStart, Nairobi, Kenya, and is Executive Chairman of Wanda Organic (www.wandaorganic.com).
Crescent Road, Kalson Towers, 6th Floor, PO Box 64142, 00620 Nairobi, Kenya.
nick.moon@kickstart.org

South Africa: an early adopter of GM crops

Jennifer A. Thomson

South Africa led the way among countries in the African continent to introduce genetically modified cotton (*Bt* cotton). There are a number of reasons for this, including the early role of an interim biosafety regulatory body, the formation of an organisation, AfricaBio, which aided information dissemination on GM crops, and finally the fact that South Africa is home to many highly sophisticated commercial farmers.

South African Committee for Genetic Experimentation

This body – SAGENE – was established by the South African Council for Scientific and Industrial Research (CSIR), which was then the research-grant-giving agency of the government. The CSIR did this to ensure that research grants would comply with the US National Institutes of Health (NIH) 1976 Guidelines for research involving recombinant DNA molecules. One of the beneficial outcomes of this was that before a scientist could apply to the CSIR for research funding in this field SAGENE had to approve the laboratories in question as being compliant with the NIH Guidelines. As many universities in the country were eager to foster this type of research, they were forced to upgrade and equip laboratories to a given standard. The scientists in question also had to give evidence of having been trained in the correct safety standards. This approach led to a number of training courses on genetic modification and biosafety being held in South Africa, resulting in a network of scientists working on a variety of projects using genetically modified organisms (GMOs). This stimulated the growth of modern biotechnology in the country.

After some years, members of SAGENE felt it had accomplished much of what it was set up to achieve and therefore went into abeyance during the late 1980s, although it continued to meet from time to time. However, this semi-retirement was to change radically with the advent of GM crops. In 1990 an application was received from the multinational company, Calgene Inc., for field trials of GM cotton resistant to the herbicide, bromoxynil (BXMTM). These trials were permitted by SAGENE following guidelines and regulations that were applicable in the USA and authorised by the South African Department of Agriculture in terms of the South African Agricultural Pests Control Act.

Shortly thereafter an application by Clark Cotton to conduct a US 'winter nursery production' of Bollgard® cotton seed in South Africa was also approved. Faced with these applications the government officially reconstituted the SAGENE committee, announcing it in the *Government Gazette* of 15 May 1992. This allowed SAGENE to represent the interests of the public and not only the scientific community. In addition, it could advise *mero motu* (by free will) and not only on request, and

... scientists had to give evidence of having been trained in the correct safety standards

was empowered to liaise with relevant international groups concerned with biotechnology. Furthermore, it could advise on legislation or controls with respect to importation or environmental release of organisms with re-combinant DNA. The terms of reference were amended on 14 January 1994 to broaden the scope from 'organisms with recombinant DNA' to 'genetically modified organisms', the term used internationally. In March 1996 SAGENE published its *Guidelines and Procedures for Work with Genetically Modified Organisms*. The document contained two questionnaires, one for the trial release of GMOs, which included field trials for GM crops, and one for the general release of GM plants.

Monsanto was the first company to apply for general release of *Bt* cotton and SAGENE recommended that they investigate the value of the new varieties to smallholder and subsistence farmers. In 1997 Monsanto managed to convince four farmers in the Makhatini Flats region of northern KwaZulu-Natal to plant some of their *Bt* cotton seeds. At the end of that season the farmers' results were sufficiently impressive to convince more than 70 more farmers to plant *Bt* cotton. The next year over 600 followed suit and by 2010 almost all farmers in the cotton-growing regions of KwaZulu-Natal planted *Bt* cotton.

However, 1997 saw the beginning of the end of SAGENE as the Genetically Modified Organisms Act was published in the *Government Gazette* on 23 May of that year. The Act could not be implemented until the Regulations were approved and thus it only came into effect on 1 December 1999.

To give an idea of the number of applications SAGENE was handling in 1997, of a total of 27, there were 13 for maize, four for cotton, two for soya, one each for canola, strawberry, eucalyptus and apple, and four for microorganisms. The applicants included companies such as Carnia, Calgene, Infruitec (local),

Rhone Poulenc, Pannar (local), Monsanto, Delta and Pine Land, Novartis, AgrEvo and Pioneer Hybrid International (personal communication, Muffy Koch). The traits being tested included insect resistance, fungal and viral resistance and herbicide resistance.

AfricaBio

The second reason why South Africa was an early adopter came from the activities of an organisation called AfricaBio.[1] Although it was started by Professor Jocelyn Webster, a member of SAGENE, in 1999, three years after the initial adoption of the first GM crop (*Bt* cotton) by smallholder farmers in South Africa, it played an extremely important role in the subsequent enthusiastic uptake of this and other GM crops by both small-scale and commercial farmers. AfricaBio is an independent, non-profit biotechnology stakeholders' association aimed at educating government officials, regulatory authorities, the media and the public at large about agricultural biotechnology. Its membership included academics, farmer organisations, grain traders, biotechnology companies, seed companies, food manufacturers and retailers, and consumers. AfricaBio was officially registered as a non-profit, Section 21 Company in 2000. Over the years it has proven its worth as a provider of accurate and objective infor-mation on biotechnology to consumers, the media and decision makers. It has provided a regular forum for exchange of information not only between South Africans but between people from many South African Development Community (SADC) countries. In addition, AfricaBio assisted other countries such as Kenya and Malawi to develop their own biotechnology stakeholder organisations. Their workshops, which provided information and training to stakeholders from coun-tries such as Malawi, Namibia, Zimbabwe and Mozambique, have

... farmers are sufficiently sophisticated to be willing to test the latest technologies

67

been particularly successful. They have also run training and advice programmes for small-scale farmers interested in planting GM crops.

Over the years AfricaBio has put out a series of position papers on such issues as GM and biodiversity, the impact of GM on biodiversity, bioethics, intellectual property rights and farmers' rights, and GM impacts on sustainable agriculture. These and other publications provided farmers with invaluable information on the role GM crops played in various aspects of agriculture and food security both within South Africa and beyond. Later booklets were produced on *Agricultural Biotechnology: Facts for Decision Makers* and *Biotechnology: Biosafety, Food Safety and Food Aid*. For many years they have sent out monthly newsletters called *BioLines*, later *GMO Indaba* and more recently *GMO Insight*, which are quick guides on topical issues.

A few examples:
* *The Impact of Biotechnology on Africa in the 21st Century* (June 2001 – a meeting held in preparation for the World Summit held in Johannesburg in September 2002)
* *Zambia Launches its First Biotech Outreach Society* (July 2003 – and they're still working on it)
* *SA GMO Maize Crops Set to Grow* (April 2004 – and they are still growing)
* *Tanzania Jumps on GM Bandwagon – Agricultural Ministry Says They Cannot Afford to Be Left Behind* (March 2005 – but it seems they are)
* *Kenyan Minister Asks Journalists to Highlight Biotech Benefits* (June 2006 – and some of them got it right)
* Bt *Toxin Resistance: An Evolutionary Action* (March 2008 – a cautionary note on responsible stewardship of the new technology)
* Bt *Awareness Campaign for Kenya Launched plus Kenya Approves GMO Bill* (April 2009 – Kenya making great strides forward)

- *Consumer Protection Regulation Effective October 2011* (October 2011 – all food in South Africa containing more than 5 per cent GMO ingredients to be labelled)
- *AfricaBio and Partners Host Successful IRM Workshop* (December 2011 – ways to prevent insects from developing resistance to the *Bt* toxin)

Commercial farmers

The third reason is the presence of many highly sophisticated commercial farmers, not found anywhere else in Africa. These farmers are sufficiently sophisticated to be willing to test the latest technologies. As many of them export internationally they need to remain competitive. They are therefore in a position to evaluate the latest scientific developments and compare their advantages with what is already available. As a result South Africa is currently number nine in the ranks of adopters of the technology, with more than 70 per cent of the maize crop totalling some 1.8 million hectares planted.[2]

References
1   www.africabio.org
2   James C. (2011) *Global Status of Commercialized Biotech/GM Crops: 2011*, ISAAA Brief No. 43. Ithaca, NY: ISAAA.

Professor Jennifer A. Thomson is Emeritus Professor in the Department of Molecular and Cell Biology at the University of Cape Town.
Department of Molecular and Cell Biology, University of Cape Town. Private Bag X3, Rondebosch 7701, South Africa. jennifer.thomson@uct.ac.za

Biotechnology and small-scale farmers: an industry viewpoint

Jorgen Schytte/Still Pics/Specialist Stock

Julian Little

Worldwide, 1.4 billion people live in poverty – of whom 1 billion are in rural areas. The problem is particularly acute in rural sub-Saharan Africa, where more than 60 per cent of the rural population experience conditions of poverty. Recent reports show that issues of poverty can be best tackled by investment in the agricultural sector, with GDP growth in agriculture contributing twice as much to poverty reduction as any other sector.[1,2] From a global perspective, the combination of enhanced productivity and efficiency generated by GM technology already provides a major boost to farmer income. Between 1996 and 2009 it was the equivalent of adding over 4 per cent to the economic value of global production of the four main crops of soya beans, corn, cotton and canola.[3]

About 16 million farmers grow over 160 million hectares of GM crops in 29 different countries according to figures published in 2012, and over 90 per cent of these were resource-poor farmers.[4] Future projections made by the agricultural biotechnology industry indicate that advances in GM technology will have particular relevance for areas where drought is a common occurrence and access to irrigation is limited. Commercialisation of drought-tolerance technology, which allows crops to withstand periods of low soil moisture, is anticipated within five years.

Just under half of the land grown with GM varieties is in developing countries[4] – an area equivalent to the surface area of Ghana, the Ivory Coast and Burkina Faso put together. In the African continent commercialised GM crops include maize, cotton and soya beans, with the number and diversity of crops increasing all the time. Trials are currently in progress on sorghum, bananas and cassava, while other developing countries grow GM squash, papaya, tomato, sweet peppers and oilseed rape. Resource-poor farmers report that the technology increases yields through greater pest and disease resistance, and this results in lower machinery and fuel costs. But it also has other benefits.

More efficient land use and food security

The amount of arable land available for agriculture worldwide is declining, especially in the developing world. Research from the United Nations estimates that more than 70 per cent of land available in sub-Saharan Africa and Latin America already suffers from severe soil and terrain constraints. With a growing population, there is little doubt that crop productivity has to increase. Unsurprisingly, the UN estimates that 80 per cent of the required rise in food production between 2015 and 2030 will have to come from intensification in the form of yield increases and higher cropping intensities.

About 16 million farmers grow over 160 million hectares of GM crops in 29 different countries

71

The amount of arable land available for agriculture worldwide is declining Productivity gains from the application of industrial biotechnology in agriculture have had a big impact on its ability to keep pace with global demand for commodities. If such crops had not been available to farmers in 2009, maintaining global production would have required additional plantings of nearly 3.8 million hectares of soya beans, nearly 6 million hectares of maize, nearly 3 million hectares of cotton and 0.3 million hectares of canola. Between 1996 and 2009, 229 million tons of additional food, feed and fibre were produced thanks to the use of GM crops. Without this, it is estimated that an additional 75 million hectares of conventional crops would have been required to produce the same tonnage.[3] Some of these additional hectares could have required fragile marginal lands, which are not suitable for crop production, to be ploughed and for tropical forest, rich in biodiversity, to be felled.

Practical benefits to farmers

Some of the benefits of seed technology uptake are tangible; others are aspirational. For example, for 80,000 farmers in Burkina Faso working an average of 3 hectares, the advent of GM cotton has meant a huge reduction in the existing use of insecticide,[5] where up to 18 sprays may be required in a particularly bad season. There has also been an immediate and substantial yield increase as well as reduction in costs, harm to the natural environment and poisoning of the farmers and local population.

In other cases, such as the development of disease-tolerant bananas in Uganda, it remains a work in progress. In central Uganda, one of the main banana-growing regions, banana *Xanthomonas* wilt (BXW) hits up to 80 per cent of farms, sometimes wiping out entire fields. To get rid of BXW, it is necessary to dig up and burn the affected plants, disinfect all machinery and tools and allow the ground to lie fallow for six months before replanting. For small-scale farmers, leaving their gardens lying empty for as long as this is not an option so they switch to other crops.

The International Institute for Tropical Agriculture (IITA) and the African Agricultural Technology Foundation (AATF) have been developing a GM solution to the problem of BXW, in conjunction with a Taiwanese biotechnology institute, Academia Sinica (AS). The institute has issued IITA and AATF with a royalty-free licence to use a new gene technology known as hypersensitive response assisting protein (HRAP). Academia Sinica successfully transplanted the sweet pepper HRAP gene into the other vegetables where it produces a protein that kills cells infected by disease-spreading bacteria. This is the first time it has been tried with a banana. Initial trials are promising, with six out of eight strains showing 100 per cent resistance to BXW. Development of wilt-resistant bananas has now progressed to the confined field-crop testing stage.[6]

Regional differences in the response to adoption of GM technologies require close scrutiny because the technology may not be the best solution in all situations. For example, in India, where there has been wide experience of the use of GM cotton, higher yields have been particularly beneficial for women.[7, 8, 9] Harvesting is primarily a female activity, therefore the women hired to pick the increased production have seen increases of 55 per cent in average income, equivalent to about 424 million additional days of employment for female earners across the whole Indian crop.

GM should not be ignored as a tool for ensuring greater food security and reliability of agricultural supply

There have been complaints from some farmers in Maharashtra that the seeds have not improved yield or met expectations of resistance to pests and

73

Seed technologies offer solutions and opportunities to small-scale farmers to improve rather than change fundamentally what they are doing diseases. Some campaign groups have interpreted this as a cause of increased suicides among farmers who have found themselves sinking deeper into debt. Yet investigations into any increase in suicide rates have suggested little or no correlation with the use of GM cotton; there have also been increases in the costs of fertiliser, pesticides and other farming supplies together with the effects of years of drought.[10] Clearly, rigorous analysis of research is therefore essential to ensure that the technology is not being oversold and that it is being adopted in the right circumstances and environments.

Investment models that work at all scales

Research on GM crops is currently thriving in Africa, with public–private partnerships looking at everything from disease-resistant bananas to drought-resistant sorghum. But for many crops, there is no obvious payback and an alternative business model is required; quite simply, how do companies overcome the cost of developing a new product when there is little chance of recuperating costs?

If it is a commodity crop such as cotton, technology will have already been developed or partially developed for markets elsewhere in the world, with R&D costs recovered through increased seed prices. The costs of this are high, and the industry's top ten companies invest US$2.25 billion, or 7.5 per cent of sales, in R&D and innovation.[11] But the resultant GM seeds are priced appropriately for each market where they are sold. They may be more expensive than conventional seeds but the resulting savings and higher income potential make them a good investment for resource-poor farmers, through lower pesticide and herbicide costs and more reliable and higher yields.

In other cases, crop traits required for particular environmental, economic and political conditions may not be applicable on a global scale and therefore will not attract the same model of commercial investment. Such projects could not proceed without both investment and an understanding that the payback period might be extremely long. In effect it requires the establishment of public–private partnerships in which companies waive or limit their intellectual property rights to the use of specific genes and transformation techniques, allowing the benefits of this technology to be maximised. Two examples include the Water Efficient Maize for Africa (WEMA) partnership, and the development of biofortified, drought-tolerant sorghum which are described in the essays of Denis Kyetere and colleagues (pp. 51–57) and Florence Wambugu (pp. 45–50).

Conclusion

GM is not a silver bullet, but it should not be ignored as a tool for ensuring greater food security and reliability of agricultural supply in Africa. Seed technologies offer solutions and opportunities to small-scale farmers to improve rather than change fundamentally what they are doing. By meeting their subsistence needs and improving the standard of living of their households, the extra income can increase the purchasing power of farmers and promote local, regional and national economic growth.

References
1   Lin L., Mc Kenzie V., Piesse J., Thirtle C. (2001) *Agricultural Productivity and Poverty in Developing Countries*, Extensions to DFID Report No. 7946. London: DFID.
2   Schneider K., Gugerty M. K. (2011) Agricultural productivity and poverty reduction: linkages and pathways. *The Evans School Review* **1**: 56–74.
3   Brookes G., Barfoot P. (2011). *GM Crops: Global Socio-Economic and Environmental Impacts 1996–2009*. Dorchester, UK: PG Economics Ltd.
4   James C. (2011) *Global Status of Commercialized Biotech/GM Crops: 2011*, ISAAA Brief No. 43. Ithaca, NY: ISAA.

5   Vitale J., Glick H., Greenplate J., Traore O. (2008) The economic impacts of second generation *Bt* cotton in West Africa: empirical evidence from Burkina Faso. *International Journal of Biotechnology* **10**: 167–183.

6   Academica Sinica: www.sinica.edu.tw/manage/gatenews/showsingle.php?_op= ?rid:404326isEnglish:1

7   Choudhary B., Gaur K. (2011) *Socio-Economic and Farm Level Impact of* Bt *Cotton in India*. New Delhi, India: ISAAA Biotech Information Centre.

8   Special issue on GM crops and gender issues: *Nature Biotechnology* **28**, July 2010. www2.warwick.ac.uk/newsandevents/pressreleases/gm_crop_produces

9   Kathage J., Qaim M. (2012) Economic impacts and impact dynamics of *Bt* (*Bacillus thuringiensis*) cotton in India. *Proceedings of the National Academy of Sciences of the United States of America* **109**: 11652–6.

10   www.tandfonline.com/doi/abs/10.1080/00220388.2010.492863

11   Crop Life: www.croplife.org/intellectual_property

Dr Julian Little *is UK Communications and Government Affairs Manager of Bayer CropScience.*
julian.little@bayer.com

Private-sector R&D, supply chains and the small farmer

Marco Ferroni

The United Nations Development Programme (UNDP) *Africa Human Development Report 2012* makes it starkly clear that for Africa to realise its long-term potential it must boost agricultural productivity. This essay points to some preconditions for success. Productivity growth requires the involvement of the private sector at all stages of the 'farm-to-fork' supply chain, starting with the research phase where innovations such as improved seed and other productive resources are progressed.

African agriculture has an uneven record of being able to feed its people, including farmers themselves. But I am optimistic. A decade of impressive economic growth has fostered change. Coupled with a shift in expectations, this bodes well for an agricultural transformation that could lift productivity, rural incomes and livelihoods in new ways. The burden of poverty, disease and malnutrition has not disappeared. But the prospects for agriculture, rural development and food security have improved with greater government commitment under the New Partnership for Africa's Development (NEPAD) programmes, rising commodity prices, and a reduction in agricultural taxation relative to the past.

The picture is one of openings in my view: Africa is full of opportunities waiting to be seized by hard-working farmers and investors willing to take on primary production risks. Three features shape the scene: the productivity gap in agriculture, unprecedented domestic and international demand growth, and breakthrough technologies. Symbolic of the latter are the mobile phones and applications used productively by ever-larger numbers of Africans, including small-scale farmers.

The scope for agricultural intensification and value addition in the food supply chain is huge. I expect returns on investment in farming to rise going forward. Small family farms are included in this dynamic; many examples show how they can profit with adequate mentoring and support. Worsening physical con-straints such as soil depletion and climate change need to be tackled. But the need to feed people and the prospect of rewards from commercial agri-culture for domestic and foreign markets – the most powerful form of agricultural development and growth – will spur innovation to address these challenges. Adaptation, gains in resilience, intensification and diversification will all play a role.

Productivity growth requires the involvement of the private sector at all stages of the 'farm-to-fork' supply chain

To see how this can work, and to understand the role of the private sector and how to nudge it forward, it helps to ask what farmers want. Irrespective of gender, age, farm size or other characteristics, their usual answer is *'technology, services and access to markets'.* The private sector supplies much of this now and will contribute more in the future.

Africa is full of opportunities waiting to be seized by hard-working farmers and investors

Technology revolves around genetics and plant breeding, soil fertility solutions, crop protection, irrigation, labour-saving devices and connectivity. *Services* include advice ('agricultural extension'), better organisation to use innovation and increase farmers' political and market clout, as well as financial services such as credit and crop insurance. *Access to markets* is about the ability to sell crops and other products and buy off-farm inputs. In good measure, this is con-ditioned by transport, storage infrastructure, information and logistics, which Africa must improve.

Governments have a big responsibility in agriculture. They must create sup-portive environments for farmers and related businesses. They should invest in capacity-building, research and the dissemination of research results. But they are generally poor at marketing farm produce and transferring inputs, services and products to farmers. These are private-sector tasks. Governments can support them through partnerships until they are viable alone. The private sector is already driving progress in key agricultural technologies, including provi-sion of fertiliser, mechanisation, irrigation, agrochemistry and – back to mobile phones – connectivity. On the question of improved seed, please see below.

Strong links to markets are essential for investment in productivity growth on farms to make financial sense. African output markets are dynamic. Consumer demand for more and better goods is strong at home and abroad, creating opportunities for farmers and businesses along the supply chain. Kenyan

Breeding in Africa is intrinsically difficult because of the diversity in production patterns and the large number of different crops and agro-ecological conditions

smallholders earn good money from 'flying vegetables' sold in UK super-markets, for example. 'Cassava+', a partnership between the Dutch company DADTCO and the International Fertilizer Development Center, builds on the growing demand across Africa for cassava by-products. Cassava+ aims to transform cassava from a 'subsistence' food to an income source for millions of small farmers by developing industrial markets and building supply chains. Yield-raising inputs and practices will also improve local food supply.

Crop improvement research and improved seed are both crucial to help raise farm productivity. Plant breeding is of overriding importance, but to make investment in it worthwhile, farmers must have access to the resulting seed. This requires seed systems, i.e. formal arrangements for multiplication, quality management and certification, and distribution or commercialisation. Seed systems are not yet well developed in most of Africa. Planting material is of low quality. Breeding itself is under-resourced and requires significantly increased public investment.[1]

Plant breeding in Africa is largely a public-sector activity, with private breeding focusing on some hybrid (where seed cannot be saved from year to year) and high-value crops such as maize and certain vegetables. Breeding in the public sector has long been affected by funding shortfalls and institutional challenges. Breeding in Africa is intrinsically difficult because of the diversity in production patterns and the large number of different crops and agro-ecological conditions encountered there. Some progress has been made and many varieties have been released. Distribution to farmers has not worked well.

Partnerships with private seed companies can greatly improve key crops. The '3G' project led by the International Potato Center with US Agency for International Development (USAID) funding, to mention but one example, has substantially increased the availability of high-quality potato seed in Kenya, Rwanda and Uganda. Viable private-sector-led seed systems have emerged in these instances, providing seed potatoes of new varieties and significantly increasing the incomes of at least 40,000 small-scale producers so far.[2] Tanzania is now replicating the model, largely with private investment.

The breeding and seed systems problem is in part conditioned by the 'lie of the land' in terms of capability and incentives, with the private sector displaying limited interest where hybrid technology does not apply and the public sector focusing on 'knowledge development' and 'public goods' as different from products and their dissemination to farmers. Partnerships appear again to be at least part of the answer, combining the skills and comparative advantages of private and public actors while at the same time promoting the use of state-of-the-art knowledge and technology, including molecular plant science and biotechnology. 'Molecular breeding' may be defined as the identification and use of associations between genotypic (DNA sequences) and phenotypic (plant trait attributes) variation to select and assemble traits into new crop varieties. A growing number of public breeders in Africa use molecular methods now, making this an increasingly dynamic area in conjunction with entities with genotyping capability, such as BecA (the Biosciences eastern and central Africa Hub) in Nairobi (see pp. 33–39) and public–private partnerships such as the African Orphan Crops Consortium.

The need to raise productivity in the context of food needs and growing markets will lead to an expansion of

Partnerships with private seed companies can greatly improve key crops

breeding by the private and public sectors as well as novel forms of cooperation between the two. Seed and input markets will become 'deeper', delivering value to farmers. It is a self-repeating, interactive chain: technology, services and access to markets.

References
1   Lynam J. (2011) Plant breeding in sub-Saharan Africa in an era of donor dependence. *Institute of Development Studies Bulletin* **42**: 36–47.
2   Kayuongo W.G. *et al.* (2010) *Effect of Field Multiplication Generation on Seed Potato Quality in Kenya*, paper presented at the 1st Triennial Meeting of Agronomy and Physiology Section (AgroPhysiology-2010) of the European Association for Potato Research, Nevsehir, Turkey, September 2010.

Dr Marco Ferroni *is Executive Director, Syngenta Foundation for Sustainable Agriculture, Basel, Switzerland.*
Syngenta Foundation for Sustainable Agriculture, WRO-1002.11.52, Postfach CH-4002, Basel, Switzerland. marco.ferroni@syngenta.com

SABIMA: an initiative for safe and high-quality GM crops

Dominique Halleux/Biosphoto

Walter S. Alhassan

B iotechnology represents a powerful tool that augments conventional approaches to tackling the future challenge of food security. The global adoption of GM technologies in agriculture has seen phenomenal growth from 1.7 million to 160 million hectares over the period 1996–2011. Genetically modified (GM) crops have delivered substantial agronomic, environmental, health and social benefits to society at large,[1] but the rate of GM growth in Africa is low, even though it is a continent faced with food insecurity. The rate of growth of GM crops is even lower in Europe but then Europe does not presently face a comparable food security challenge.

Safety concerns

Genetic modification enables the transfer of genes artificially from one organism to another for a specific purpose (e.g. increased yield and nutritional value, protection against pests and diseases, survival in hostile environments). In this way hereditary material (genes) can be moved to or from unrelated species in a controlled and predictable manner that supplements and extends the normal process of plant breeding. Currently the following GM crops have been commercialised: maize, soya bean, cotton, canola (rape), squash, papaya, sugar beet, tomato, sweet pepper and alfalfa. They are grown primarily in North and South America, South and East Asia.

Production of a GM crop passes through the following stages, each being subjected to stringent scrutiny: (1) laboratory, (2) containment greenhouse, (3) field-testing under strictly controlled conditions ('confinement'), (4) extensive risk assessment (multilocational trials including farmer fields, food safety testing), (5) farm release and (6) post-market surveillance. This process, including regulatory approval, is very costly, which means that in the main it is restricted to a few staple crops and can only be afforded by private, multinational companies and not publicly funded research organisations.

Science academies the world over, government studies, United Nations agencies and religious bodies have not found any new risks associated with GM crops used for human food or animal feed that have undergone the required safety checks before release. As Sir John Beddington, UK Government Chief Scientific Adviser pointed out recently, over a trillion meals have been made using GM crops in North America and there has not been a single case in the law courts of anyone suing after eating GM products.[2]

An underlying factor for the slow growth of GM crops in some continents remains scepticism over their safety to humans and the environment despite repeated assurances from respected independent bodies. In 1999 Britain's Nuffield Council on Bioethics reported: *'We have not been able to find any evidence of harm. We are satisfied that all products currently on the market have been rigorously screened by the regulatory authorities, that they continue to be monitored, and that no evidence of harm has been detected'.*[3] National academies of science in France, Germany and the UK have since reached similar conclusions (2002 and 2003). The Food and Agriculture Organization of the United Nations (FAO) said that *'to date no verifiable untoward toxic or nutritionally deleterious effects resulting from the consumption of foods derived from genetically modified foods have been discovered anywhere in the world'.*[4] Scientists in New Zealand and The Netherlands surveyed worldwide literature and concluded that GM crops had been no more invasive or persistent than conventional crops and were not more likely to lead to gene transfer (2003).

An underlying factor for the slow growth of GM crops in some continents remains scepticism over their safety to humans and the environment

More recently, a Study Week of the Pontifical Academy of Sciences in Rome brought together distinguished academy members and other experts who reported that *'there is a moral imperative'* to make the benefits of genetic engineering technology *'available on a larger scale to poor and vulnerable populations who want them'*, urging opponents to consider the harm that withholding this technology will inflict on those who need it most (2009).

The European Commission's Directorate-General for Research and Innovation, reviewing the past 25 years, stated that biotechnology, and in particular GM crops, are no more risky than conventional plant breeding technologies.[5] Islamic scholars of modern biotechnology concluded that GM as a method of plant improvement is not intrinsically different from other plant improvement techniques from the *shariah* point of view.[6] Acceptance of GM processes and products was recommended for Muslims and members of the Organisation of the Islamic Conference as long as the sources from which they originated were *Halal*. The only restrictions were products derived from *Haram* origin that retained their original characteristics and were not substantially changed.

Stewardship and self-regulation: a new concept in African modern biotechnology product development

Recently, an African initiative has developed a self-regulating index of responsible management or *stewardship* of modern biotechnology. This form of stewardship aims to secure GM products that have proved safe and stable over the entire product cycle – from gene discovery to seed production to market place.

This project on *Strengthening Capacity for Safe Biotechnology Management in Sub-Saharan Africa* (SABIMA) was introduced in Africa by the Forum for Agricultural Research in Africa (FARA) to fill a gap in the training of African scientists to conduct modern biotechnology research, and to develop and disseminate products to ensure that they were not only meeting regulatory compliance for safety but also addressed product quality requirements. The SABIMA project focused on stewardship training and adoption in biotechnology and in GM product development as the core objectives. Stewardship is the self-regulating responsible management of a (GM) product from its inception stage through the entire product line. For each stage of the product development, an analysis is performed (called critical control point analysis) to determine what could go wrong and pre-emptive action is taken through standard operation procedures (SOPs) to address potential challenges, if any. As a first step, the product in development, if it is a GM product, must meet the

The adherence to best stewardship practices enjoins all individuals ... to practise good stewardship

regulatory requirement for safety before the quality assurance steps are taken at each stage of the product development line.

Adherence to best stewardship practices enjoins all individuals, from the researcher, the farmer, processor, distribution/shipping agent for seed or grain to those responsible for product discontinuation and the relaunch of new products, to practise good stewardship at their point in the product cycle to ensure product integrity or wholesomeness. Lapses in adherence to stewardship guidelines have created problems in world trade through the mixing of GM and non-GM product due to improper product segregation or separation during the handling of merchandise. Challenges can also arise in the management of GM products on the farm by not adhering to mandatory buffer zones of non-GM plants (refugia) around GM crops. Failure to meet this requirement encourages the development of resistant insect pests to the introduced gene, and such resistance can present a growing global problem.

So far six countries in sub-Saharan Africa (Burkina Faso, Ghana and Nigeria in West Africa; Kenya and Uganda in East Africa and Malawi in Southern Africa) have benefited from the SABIMA project. Thirteen FARA-certified trainers have been produced by the project and they have in turn trained 1,412 people in the six project countries. Case studies on various management challenges identified and addressed through stewardship principles learnt have been published by FARA and made available at its website.[7] One such stewardship application was in the development of GM (*Bt*) cotton in Burkina Faso by scientists at the Institute for Environmental and Agricultural Research (INERA), Ouagadougou. Here, the challenge was in avoiding the mixing of *Bt* (GM) and non-*Bt* (non-GM) cotton by farmers who must handle both crops on the same farm as well as ginnery operators who must also handle both products. Farmers, extension officers and ginnery operators had to be trained in stewardship principles by INERA scientists who had themselves been trained in stewardship principles involving proper packaging and labelling, cleaning to avoid residual contamination and batch handling of GM and non-GM cotton at planting and at the ginnery. These are listed critical control points whose analysis determines steps to be taken to avoid the mixing of *Bt* and conventional non-*Bt* cotton. These interventions by INERA have reduced the incidence of cotton seed mixing.

In Malawi, through critical control point analysis by SABIMA project-trained stewards at the Chitedze Research Station, Lilongwe, Malawi, harvest of certified maize hybrid seed was detected as a critical point for detecting farmer adulteration of certified seed with ordinary grain. Acceptable yield levels for certified seed from farmers trained on the importance of product stewardship diminished the unwholesome practice of certified seed-grain mixing. The Malawi case is an example of the use of stewardship principles in a non-GM crop setting.

At the First Pan-African Conference on Stewardship in Agricultural Biotechnology in Ghana on 29–30 November 2011,[8] there was not only an overwhelming endorsement of the SABIMA project but a call for its outscaling to a minimum of ten more countries. FARA is the first and so far the only institution in Africa addressing training in the crucial area of stewardship in biotechnology product development and complementing other organisations addressing solely bio-safety capacity-building and awareness creation/advocacy for biotechnology. FARA is currently in search of funding support to outscale the SABIMA project that ended in 2011 into a second phase that will also see its mainstreaming into university curricula.

Concluding thoughts

Despite 15 years of successful GM crop commercialisation globally, there has not been a scientifically proven adverse effect on human and animal health and the environment attributable to its release. The growing evidence is that both GM and non-GM food crops are substantially equivalent. Nevertheless GM crops are still subject to strict regulatory compliance in their development.

Stewardship, a self-regulating requirement, ensures both product safety and quality and indeed the sustainability of modern biotechnology. The SABIMA project is designed to build a culture of stewardship in GM and non-GM product development on a continuing basis. African countries are encouraged to link with private-sector organisations in crop biotechnology stewardship to continuously benefit from new trends in biotechnology stewardship. FARA's role in the maintenance of a community of practice to share stewardship best practices in Africa will be crucial.

There is a sense of growing optimism that over the next five to ten years there will be acceleration in the pace of development and use of GM products on a need basis for Africa's food security. Reasons for this include the fact that there is a pipeline of GM food products being developed in Africa by African scientists in public institutions and not by biotech companies. The food products (at confined field-trial stage) are the ones that will be largely consumed by Africans and will include *Bt* cowpea, high carotene banana, biofortified sorghum, nutrient-enhanced cassava and sweet potato, water-efficient maize and nutrient-efficient rice. The African countries involved in the research include Burkina Faso, Egypt, Ghana, Kenya, Nigeria, South Africa and Uganda. The wholly consumed African products will hit the market under country-specific biosafety regulations. Efforts are ongoing to harmonise the regulations over various sub-regions. Continuous awareness creation, transparency, investment in research and farmer support services are among the needed catalysts.

References
1   James C. (2011) *Global Status of Commercialized Biotech/GM Crops: 2011*, ISAAA Brief No. 43. Ithaca, NY: ISAAA.
2   Combating global food crisis with GM crops. EfeedLink. 30 January 2012. www.efeedlink.com/contents/01-30-2012
3   www.nuffieldbioethics.org/gm-crops
4   www.fao.org/docrep/015/i2490e/i2490e04d.pdf
5   http://ec.europa.eu/research/biosociety/pdf
6   www.worldhalalforum.org
7   www.fara-africa.org
8   Report at www.fara-africa.org/media/uploads/stewardship_conf_report_ver03_whole_book.pdf

Professor Walter S. Alhassan is a consultant on biotechnology and biosafety policy issues for the Forum for Agricultural Research in Africa (FARA) in Accra, Ghana.
FARA Headquarters, 12 Anmeda Street, Roman Ridge, PMB CT 173, Cantonments, Accra, Ghana. walhassan@fara-africa.org

Preparing youth for high-tech agriculture

Ron Giling/Lineair/Still Pictures

Margaret Karembu

A prediction from international circles is that in the next 50 years the world will need to produce as much food as has been produced in the history of mankind. The world population is expected to reach 9 billion by then, translating to the highest demand for food ever. The irony is that few people can provide convincing responses to what I would consider the million-dollar questions: Where will the food come from? How will it be produced? Who will produce it? Depending on the region one comes from, some will say the food will come from supermarkets; for others, organic farming, intensification of agriculture, modern technologies, distribution from the 'haves' to the 'have-nots'; while others will just wish you away. What is certain is that food production will have to double or triple using fewer resources. With

Global agricultural and food systems will have to change substantially to meet the challenge of feeding the expanding population

additional challenges from climate change and dietary shifts to greater meat consumption, global agricultural and food systems will have to change substantially to meet the challenge of feeding the expanding population. This will be more daunting in Africa where the population is projected to have doubled by then.

Where will the food come from?

A number of proposals have been advanced on potential global food sources, ranging from opening up new land to enhancing access to inputs and improving food distribution across the world. Common knowledge, however, asserts that most of the land suitable for agriculture in Africa has been overexploited while food is grown and consumed at source with very diverse cultural preferences. Encroaching into marginal lands, traditionally reserved for wildlife, has made a bad situation worse, resulting in loss of valuable biodiversity and in human–animal conflicts. Yet, this is the same land expected to feed the expanding population that is getting poorer and increasingly food-insecure by the day. Producing enough food for all will therefore require a 'business unusual' model – one that utilises the best of conventional methods with the best of appropriate technologies that demand the least amount of land with the least aggregate of external inputs in extremely harsh conditions (drought, salinity, flooding, etc.). Clearly, an appropriate use of modern scientifically based agricultural technologies will be essential.

How will the food be produced?

In a meeting of minds in agriculture in September 2012, the African Green Revolution Forum declared that Africa's agriculture remains backward due to limited application of modern science and technology. Despite being a source of livelihood for about 70 per cent of the population, the continent is still overdependent on rain-fed agriculture. It is also marked by low use of high-yielding seeds, fertilisers and pesticides. While addressing the Forum, the Tanzanian President His Excellency Jakaya Kikwete lamented: *'Farmers lack modern agricultural production skills and knowledge and do not have access to financial and other supportive services. Consequently, farm sizes are small, yields are low and*

Appropriate value chains with clearly defined service providers at each stage of the chain would also ensure that farmers do not become victims of their own success, with increased production failing to reach the market

revenues from the agricultural sector remain meagre.' Importantly, increasing productivity per unit of land, diversification of food-eating habits, access to high-yielding varieties and markets, increased nitrogen-use efficiency, breeding for tolerance to abiotic (drought, salinity, flooding) stresses and agribusiness will be the cornerstones of revitalising African agriculture, the Forum recommended. Clearly the modern science of genetics will have a key role to play.

Other land-saving technologies such as hydroponics and sunlight greenhouse farming where farmers can control and dictate their seasons without having to rely on rainfall patterns are necessary. They will, however, require aggressive promotion and support with high-yielding seed varieties, links to markets and guaranteed access to credit and inputs. Appropriate value chains with clearly defined service providers at each stage of the chain would also ensure that farmers do not become victims of their own success, with increased production failing to reach the market.

Who will produce the food?
A typical characteristic of African agriculture is the predominance of small-scale farmers (more than 60 per cent), mostly women who are semi-skilled and with little access to technology, inputs, credit and markets. Migration of young

people from rural to urban areas has left food production in the hands of their elderly parents, most of whom are incapable of adjusting to modern high-tech farming systems. Farmers have therefore remained at the level of mere subsistence, with little mechanisation and rudimentary farming methods, a situation that perpetuates poverty and food insecurity. The status quo has only served to further demotivate the youth as farming is portrayed as a punitive, inferior and non-profitable enterprise. In addition, young people do not view themselves as part of the solution to the food insecurity problem. Yet their population is increasing at an alarming rate, higher than that of economic growth. The 2009 Population Reference Bureau report projects that there will be about 343 million young people (defined as people between the ages of 10 and 24) by 2015 – a big labour force with the potential to contribute to feeding the world!

How can youth be attracted and retained in the agricultural sector given their view of agriculture as a painful and low-end labour market? At the outset, youth are not being integrated in the agricultural sector, leaving food production in the hands of the elderly. The majority throng to cities in their millions, ending up in slums and on the streets doing menial jobs and hawking all manner of counterfeit imported goods – a sure way of killing their own innovations and the economy. The returns are low and, with time, many young people lose heart and some join criminal gangs or indulge in the illicit drinks trade. There is need therefore for a fundamental change in the mindsets of African youths to view themselves as key players in the food production chain.

There is need for a fundamental change in the mindsets of African youths

This can be possible if farming becomes pleasurable and profitable with supportive infrastructure to make it exciting, worthwhile and recognised as an important contribution to modern society. Access to efficient technologies and assurance of access to high-quality seeds, inputs and links to markets would be good starting points. This would be a remarkable change from what rural societies have done over the years. According to Dr James Mwangi, a Kenyan and winner of the 2012 Ernst & Young World Entrepreneur award, also a council member at the G8 New Alliance for Food Security and Nutrition, lessons can be borrowed from fast-growing sectors with successful youth entrepreneurship ventures. Examples include the Techno-entrepreneurship (ICT) and social entrepreneurship sectors using social causes (cooperatives) to drive entrepreneurship and growth.[1] Kenya and Ghana have invested in facilitating incubation centres for ICT and social entrepreneurs such as iHUB and mLABs. An ICT hub (iHUB) is a space where technologists congregate to bounce ideas around, network, work, program and design. This is achieved through an enabling environment where a community of tech entrepreneurs can grow and innovative ideas can be born from collaborations and the atmosphere of the co-working space.[2] The mLABS on the other hand are wireless informal focus group meet-ups aimed at creating forums for exchange of views and networking between mobile application developers and practitioners in various industry sectors. Simulation of the introduction of such models in the agricultural sector could provide opportunities for youth to develop successful agribusinesses.

Further, there is compelling evidence that modern biotechnology applications such as tissue culture can greatly enhance productivity by generating large quantities of disease-free, clean planting material. This is applicable to the mass production of Africa's key staple foods like banana, cassava, sweet potato and yam.[3] Youth with a first degree in agriculture or biological sciences should be encouraged and facilitated to establish low-cost tissue-culture business facilities at community level. African governments should also start considering young people's views and perceptions in policy-making and in reviewing agricultural and biosciences education curricula.

Experience of nearly two decades with biotech crops globally has demonstrated the power of marker-assisted conventional breeding and genetic engineering in developing superior crop varieties resilient to various biotic and abiotic stresses. Modern biotechnology has enabled development of crop varieties that can withstand pest attack, particularly insects and weeds, and nutritionally enriched (biofortified with vitamins and micronutrients such as iron and protein, vital for women and for the early growth and development of children). Extensive work is also ongoing to develop drought-tolerant and nutrient-use-efficient crops appropriate in Africa. Such powerful technologies will further increase opportunities for young people to pursue more efficient, pleasurable agricultural enterprises with high chances of success. Even without land, they can engage in a revitalised agricultural economy not controlled by large farms but with high-value crops and land-saving technologies such as modern biotechnology.

The myths surrounding agriculture, which portray it as punitive and an occupation for the desperate, should be challenged as efforts are directed to encouraging youth engagement in the sector. Radical measures such as the use of celebrities (musicians, artists, comedians, etc.) to spearhead food security campaigns could also ignite passion for agriculture among young people, as is happening in other social spheres. Platforms should be explored to promote successful youthful role models in agriculture through popular social media – for example, Facebook, Twitter, etc. – where young farmer entrepreneurs could share their experiences. Literature abounds with examples of small businesses of various sizes and industries that have transformed their online presence with innovative social media marketing practices.[4] By organising youth for higher-tech agriculture using social media, greater receptivity could be expected, particularly in risk aversion, mental activity and scientific understanding, since the majority of young people greatly trust social media.

Conclusion and moving forward

Agriculturalists agree that the long-term sustainability of existing food production systems will largely depend on appropriate uptake and application of modern science and technologies. Education, empowerment and motivation of young people to take up agricultural activities are a prerequisite for improved and sustainable food production in Africa given their big numbers. However, this is not an overnight endeavour and calls for long-term investment and an overhaul of agricultural education curricula and support systems that enable the youth to apply agricultural innovations in a pleasurable and profitable way. The mass media have an important role in changing this perception. With better opportunities for access to technologies, entrepreneurial skills and social marketing, young people could funnel their youthful idealism, energy and determination into a positive force for change within the agricultural sector. This would ultimately result in sustainable production of the food required to support the growing population in Africa.

References
1   http://digitalnewsagency.com/stories/6683
2   www.africalabs.com
3   www.isaaa.org/resources/publications/pocket/14/default.asp
4   www.socialmediaexaminer.com/9-small-business-social-media-success-stories

Dr Margaret Karembu is Director of the International Service for the Acquisition of Agri-biotech Applications (ISAAA) AfriCenter, based in Nairobi, Kenya.
ILRI Campus, Old Naivasha Road, Uthiru, 00605 Nairobi, Kenya. mkarembu@isaaa.org

97

Risks to biodiversity: real or imaginary?

Ghillean T. Prance

I am delighted to contribute to this collection of essays because I have long seen the potential of genetic engineering (GE) of crops as a tool to help feed the developing world. The idea that it is only more developed countries which profit from the use of genetically modified (GM) crops is increasingly shocking when we see that one in eight people in the world continue to go hungry. It is my hope that these essays will encourage all leaders to investigate the potential of genetically modified crops for their region and not be put off by the negatives and misrepresentations that have circulated so widely.

The new focus of agriculture is sustainable intensification ... a strategy in which GM crops can play their part

The new focus of agriculture is *sustainable intensification*. This is a key recommendation of a recent report from the Royal Society,[1] the UK's leading academy of science. It is a strategy for converting research into practice for agricultural practitioners, whether smallholders or large landholders, and for agribusinesses alike, and one in which GM crops can play their part. However, I have been a known critic of the first generation of GM crops which were developed and commercialised primarily by the private sector. Insufficient attention was paid in the early days to their potential impact on the environment, or to addressing the needs of small farmers in African and other developing countries, many of whom could benefit from the technology if applied appropriately. This situation shows signs of change, with public efforts directed towards the conservation of biodiversity, preservation of traditional crops and landraces, and increased focus on plant genetic resources such as indigenous or 'orphan' crops. Private–public partnerships are also beginning to bear fruit with the provision of technological and financial services for smallholder farmers, which can improve their well-being and help them to become entrepreneurs. It is notable that the vast majority (over 90 per cent) of farmers growing GM crops are small-scale farmers in the developing world.[2]

Dealing with risks

A risk often cited by opponents of genetic engineering is that to expand the use of GM crops would lead to the escape of genetically engineered genes into the natural environment with consequences for related varieties and species. Several common weeds in the USA have developed resistance to herbicides and there are claims that gene flow may have already occurred in wild relatives of maize in Mexico.[3] But the development of resistance to herbicides (or pesticides) is a fact of life both in agriculture and medicine, and in biology in general. It

One of the most important aspects of biodiversity conservation is the preservation of the habitats where wild relatives of crop species occur

99

should surprise no one. Also, the flow of genes from modern conventionally produced maize hybrids into traditional crops (landraces) is well known.[4] Landraces have not perished or been destroyed, nor have there been cultural consequences as farmers in Mexico have taken advantage of the new crops.[5] Changes take place in plant genomes anyway because they are dynamic and not static parts of the plant itself. This is just as true of landraces as it is of GM crops. Landrace varieties are the product of generations of continuous crossing and selection by farmers themselves to achieve the optimal characteristics the farmer wants.

The message is that changes have to be accommodated and this can be achieved by making mindful choices before creating GM plants in the first place, and by rigorous monitoring of the development of a GM crop through the tests required by the international regulatory and biosafety regimes, which are far more demanding than for conventional plant breeding.[6]

One of the most important aspects of biodiversity conservation is the preservation of the habitats where wild relatives of crop species occur. It is vital to decide at the outset whether these plants should be preserved from 'contamination' by engineered genes with which they can interbreed. Each introduction of a genetically modified crop needs to be assessed on a case-by-case basis, and there is no reason why they should be denied to small farmers once they have been fully tested. Most crops are grown far from their place of origin but, even so, some favour a precautionary approach. Using this approach the modified seed of sugar beet or rape would not be introduced into the UK because they can both cross with native species; similarly with rape (*Brassica napus*), which can cross with its wild relative *Brassica rapa*, the wild turnip. But 'contamination' by gene flow should not be considered a 'show-stopper' for the reasons already pointed out. Maize does not cross with native grasses of the UK and the use of GM maize is very un-likely to have an effect through the distribution of modified genes into the environment.

One of the greatest potentials of GM crops in the future is to enable the use of marginal lands, especially in such places as the arid regions of Africa

It is already evident in many places that the increased intensity of farming, whether using GM or conventionally bred crops, has caused the decline of various important species of birds, butterflies and other insects. This needs to be heeded and monitored. A UK government farm-scale study of three GM crops over four years showed that in the cases of rape and beet, insect wildlife was considerably reduced, but for maize there was no loss of biodiversity.[7] GM crops should only be used after adequate research has been carried out on the effect each crop has on wildlife. *Sustainable intensification* of farming methods as envisaged by the Royal Society report[1] may help to avoid destroying more native habitats and environments. Unexpectedly, in Argentina and Brazil where GM crops are widely grown, there has been an increase in certain insect populations due to a reduced frequency of pesticide use.

Invasive species are seen as another threat to native biodiversity, namely, the danger of GM plants having the genetic make-up to out-compete native plants. However, most cultivated plants have very different characteristics from weeds, of which aggressive species are well documented. Many crops never establish themselves in nature and rarely reseed after cultivation, and this is as true for maize as it is for soya beans. In general, there appears to be no reason to fear gene flow from GM plants to relatives that produce new weeds, but every situation must be dealt with by appropriate agronomic practices, whether it involves GM plants or not.[8] For example, a transgenic strain of the creeping bentgrass (*Agrostis stolonifera*) that was bred for golf courses has spread from

101

test plots and established itself in the wild.[9] Seeds and pollen were spread in the wind and this raises the question of whether this is the beginning of a new breed of invasives that could threaten biodiversity. Or is this an example of how special situations can arise and will need to be dealt with on a case-by-case basis?[8]

Recognising the benefits

One of the greatest potential benefits of GM crops in the future is to enable the use of marginal lands, especially in such places as the arid regions of Africa. The technique of genetic modification will also be an essential tool to create crop plants that are adapted to and can tolerate climate change. Examples of the benefits of GM crops are outlined elsewhere in this book, but in terms of helping to preserve biodiversity the greatest benefit is that it can lead to sustainable intensification of land use, and the cultivation of those marginal habitats that are of little importance for conservation. The application of GM crops that leads to more intensive use of existing croplands could reduce the need to destroy more of the natural habitats that harbour so much of the remaining biodiversity of the world. For example, salt-toleration and drought-resistance traits could enable the use of marginal land rather than destroying land currently covered by pristine habitats.

Davidson recounts the tragedy of the papaya in Thailand where a GM variety was developed with resistance to the ringspot virus that was killing the plants.[10] Greenpeace protested and dumped fruit outside the Thai Parliament to protest against the legislation that would have legalised the use of this GM papaya. This inconsiderate action has resulted in an economic loss of US$850 million in 2007 and the loss of a vitamin-B-rich source of nutrition for the Thai people.[11] The use of this GM fruit is unlikely to do any more harm to biodiversity than any other ordinary fruit crop.

Conclusion

I have outlined risks and benefits of GM crops to wildlife, recognising that many of them are no different from the introduction of any new plant variety or advanced hybrids derived from the well-established methods of conventional plant breeding.

Where GM crops are used on a large scale coupled with the use of herbicides and pesticides, they are subjected to regulations which demand that they are monitored and their effects controlled in such a way as to minimise impacts on biodiversity. Later generations of GM crops in the pipeline derived by new breeding techniques (e.g. zinc finger nuclease technology, cisgenesis and intragenesis, RNA-dependent DNA methylation) will be the subject of careful evaluation of risks and benefits to ensure that genes do not impact biodiversity. Therefore, dangers to biodiversity are controllable given adequate research and legislation and should not be used as an excuse to keep almost a billion people starving.

The challenge is to make GM varieties of crops with added benefits readily and cheaply available to the starving poor around the world. Biotechnology should be within the economic reach of poor farmers because, after all, it is the smallholder farmers of Africa (mostly women) who need to produce more food consistently and by sustainable means with little adverse effect on biodiversity.

References

1   Royal Society (2009) *Reaping the Benefits: Science and the Sustainable Intensification of Global Agriculture*. London: The Royal Society.

2   James C. (2011) *Global Status of Commercialized Biotech/GM Crops 2011*, ISAAA Brief No. 43. Ithaca, NY: ISAAA.

3   Dalton R. (2008) Modified genes spread to local maize: findings reignite debate over genetically modified crops. *Nature* **456**: 149.

4   Gressel J. (2010) Needs for and environmental risks from transgenic crops in the developing world. *New Biotechnology* **27**: 522–7.

5   Parrott W. (2010) Genetically modified myths and realities. *New Biotechnology* **27**: 545–51.

6   Ellstrand N.C. (2001) When transgenes wander, should we worry? *Plant Physiology* **125**: 1543–5.

7   Coghlan A. (ed.) (2003) Theme issue: The farm-scale evaluation of spring-sown genetically modified crops. *Philosophical Transactions of the Royal Society B* **358**: 1775–1913.

8   Raven P. (2010) Does the use of transgenic plants diminish or promote biodiversity? *New Biotechnology* **27**: 528–33.

9   Reichman J.R., Watrud L.S., Lee E.H., *et al.* (2006) Establishment of transgenic herbicide-resistant creeping bentgrass (*Agrostis stolonifera* L.) in nonagronomic habitats. *Molecular Ecology* **15**: 4243–55.

10   Davidson S.N. (2008) Forbidden fruits: transgenic papaya in Thailand. *Plant Physiology* **147**: 487–93.

11   Sriwatanapongse S., Iamsupasit N., Attathom S., Napasintuwong O. and Traxler G. (2007) *The Study of Agricultural Benefits in Thailand*. Bangkok, Thailand: Biotechnology Alliance Association.

Professor Sir Ghillean Prance FRS *was Director, Royal Botanic Gardens, Kew, UK, and is now Scientific Director and a Trustee of the Eden Project in Cornwall and Visiting Professor at Reading University, UK.*
siriain01@yahoo.co.uk

Hazards and benefits of GM crops: a case study

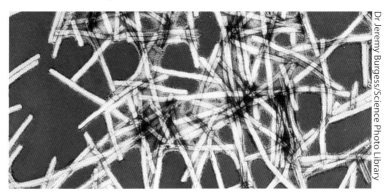

Dr Jeremy Burgess/Science Photo Library

David Baulcombe

Viruses are tiny pathogenic particles inside cells that infect other living organisms: in humans they cause chickenpox, influenza, polio, smallpox and other diseases. The first virus ever to be described infects plants – it was tobacco mosaic virus (illustrated above) – and plant viruses, like those of humans, cause disease. When they infect crops, they can be a serious problem for farmers.

Some crops are protected from viruses by disease-resistance genes. Plants carrying these genes are identified by plant breeders and refined as new varieties for use in agriculture through a lengthy crossing programme. However, the appropriate resistance genes are not always available and many crops are susceptible to virus disease. To protect these susceptible crops there is a promising new strategy that

... there is a promising new strategy that illustrates the benefits of genetic modification technologies

105

illustrates the benefits of genetic modification (GM) technologies. It also provides a new opportunity to test the level of risk of such technologies in the field.

This new GM approach has been successful in the laboratory with many viruses[1] and, in one example, with papaya, it has been used in regions of Hawaii to protect against papaya ringspot virus. This disease was previously destroying virus-susceptible plants[2] but, with the new GM varieties, the plantations have been re-established and farmers' livelihoods have been restored.

The chromosomes of these GM plants contain pieces of introduced DNA (transgenes) that include a fragment of the viral genome. Recent research indicates that these transgenes are effective in virus resistance because they reinforce a natural defence system against viruses that is known as 'ribonucleic acid silencing' (RNA silencing). Perhaps there is a message in this finding: innovation in biotechnology is best achieved by modifications to natural processes rather than by attempts to synthesise a new mechanism?

If RNA silencing is compared, metaphorically, to the immune system in humans and other mammals, the transgene that includes a piece of viral DNA is like a 'nucleic acid antigen' and the plant responds by the production of an 'RNA antibody'. In effect the foreign nucleic acid in the transgene boosts the natural defence of the crop in the same way that a vaccine protects us from polio, influenza or other viral diseases. The hope is that African crops could be protected by GM RNA silencing against maize lethal necrosis, African cassava mosaic, cassava brown streak, rice yellow mottle, groundnut rosette, banana bunchy top and many other viral diseases.[3]

... transgenes are effective in virus resistance because they reinforce a natural defence system against viruses

Of course it is not a simple matter to tackle crop disease. One of the most significant complicating factors is the ability of viruses to evolve rapidly. When we grow virus-resistant crops we introduce strong selection pressure for strains of the virus that can evade the resistance mechanism. With conventionally bred plants this problem is difficult to address and, eventually, the resistance gene is useless because the resistance-breaking viruses become so abundant. However, with RNA silencing, we can 'immunise' the crop with multiple elements of the viral DNA. This strategy would minimise the risk that the resistance is overcome because the virus would require two or more simultaneous mutations to evade the 'RNA antibody'. Mutations are rare and two simultaneous mutations at defined sites are almost impossible.

A second complicating factor is the potential for resistance genes, including transgenes, to affect the safety of the crop. A recent report, for example, suggests that plant RNA in the diet of a mammal can be taken up into the liver where it can switch off, or silence, gene expression.[4] In such a scenario the 'RNA antibody' produced in the GM plants could be hazardous if, by chance, it targets human liver genes. However, the transgene RNA would be diluted by the large amount of RNA produced naturally in the plants. There is, therefore, a much greater risk, by many orders of magnitude, from the natural plant RNAs in our diet than from the transgene. As humans eat many plants without harm it is unlikely that absorbed transgene RNA presents a hazard.

Another potential complication of GM arises if a transgene encodes a protein that affects the safety, nutritional value or quality of the crop. There is the same potential hazard with conventional breeding in which thousands of protein-coding genes with potential to cause harm are transferred into the crop from, for example, a wild relative. However, with RNA silencing, the resistance does not depend on transgene-encoded proteins. It is therefore highly unlikely that

RNA silencing would introduce a protein-based hazard to human health or the environment.

Modern agriculture uses fewer varieties of crops than traditional or local farming and there is concern that we are losing diversity in crop germplasm. The focus on few varieties is, in part, because it is difficult to transfer desirable traits by conventional breeding from a wild plant into multiple new varieties of a crop. However, this limitation does not apply with GM traits. A transgene can be introduced simultaneously into many different varieties and they would all be improved without loss of their original agronomic characteristics. To improve several varieties in this way is not a trivial undertaking but it would be much easier than with conventional breeding. A GM strategy could, therefore, preserve biodiversity in cropping systems, not reduce it.

Other hazards of RNA silencing in GM plants are similar to those associated with conventional genetic traits. It could be, just as new conventional varieties sometimes fail in large-scale trials, that RNA silencing is not as effective in the field as in the laboratory. Conversely, the GM trait could be very effective in the field and the crop could acquire the damaging invasive characteristics of weeds. However, the problem of crops as weeds is not new. In the UK, for example, the yellow flowers of rapeseed are a common sight as a weed in other crops. There is no reason to think that transgenes would be more hazardous or pose greater risk in this sense than conventional genes conferring virus resistance.

With good crop management there is no reason why virus resistance ... should not become a durable and widely used technology

A prudent approach for GM in Africa, taking these various hazards into account, would involve a testing programme similar to that used for GM crops in the UK.[5] First the trait and the potential hazards would be tested in growth chambers, glasshouses and then field plots. Progressively more extensive field trials in several locations are then used to assess the effectiveness and stability of the trait and the impact to the environment including any effects on gene flow.

However, even when virus resistance from GM is demonstrably effective and safe in the field, it should not be considered as a panacea: other protection strategies should also be used. Planting of the crop, for example, should be in rotations and at times of the year that are not compatible with the life cycle of the insects and nematodes that carry the disease from plant to plant (vectors). Similarly the weed control and tillage methods should discourage these vectors and prevent infection reservoirs that could spread to the crop. With good crop management there is no reason why virus resistance achieved by RNA silencing should not become a durable and widely used technology to help achieve food security in Africa.

References
1   Baulcombe D.C. (1996) RNA as a target and an initiator of post-transcriptional gene silencing in transgenic plants. *Plant Molecular Biology* **32**: 79–88.
2   Gonsalves D. (1998) Control of papaya ringspot virus in papaya: a case study. *Annual Review of Phytopathology* **36**: 415–37.
3   www.bis.gov.uk/assets/foresight/docs/infectious-diseases/t5_10.pdf
4   Zhang L., Hou D.X., Chen X., *et al.* (2011) Exogenous plant MIR168a specifically targets mammalian LDLRAP1: evidence of cross-kingdom regulation by microRNA. *Cell Research* **22**: 107–26.
5   www.defra.gov.uk/acre

Professor Sir David Baulcombe FRS *is the Regius Professor of Botany and Royal Society Research Professor at the Department of Plant Sciences of the University of Cambridge.* Department of Plant Sciences, Downing Street, Cambridge CB2 3EA, UK. dcb40@cam.ac.uk

Do patents hold up progress in food security?

Sean Butler

Sometimes straight questions deserve straight answers: patents do not hold up progress in food security. Progress in food security will come from improvements in many areas, including infrastructure, governance, the regulatory regime, the rule of law and the local and global economy – and also from advances in agriculture, in agronomy, in fertiliser and in seed, many of which have been improved significantly because of innovations and inventions in which the patent system plays an important enabling role.

Patents are part of the wider field of 'intellectual property', a branch of the law that recognises works of intellectual effort, and grants limited ownership for that effort. It exists to protect inventions through patents, plant varieties through plant variety rights (PVR, sometimes called plant breeders' rights, PBR), and music and literature through copyright. In the case of patents and PVR they represent a contract between society and an inventor, under which an inventor or breeder is given an exclusive right to produce their invention (plant variety) for some 20 years. In return society benefits from having access to information about the invention itself, the opportunity to use it with consent,

Patents are part of the wider field of 'intellectual property', a branch of the law that recognises works of intellectual effort, and grants limited ownership

and the right to develop new inventions by inventing around it. The result is that successful inventions can be commercialised profitably.

A basic principle of business is that commercial activity is unlikely in the absence of the opportunity for commercial gain. The financial rewards from patents in turn both deliver commercial gain and provide a commercial rationale for investment to develop more inventions. Especially in R&D-intensive industries, of which plant breeding and plant biotechnology is certainly one, the cost of the research needs to be recovered in the price of the commercial product. By obtaining patents on a product a company is able both to charge a premium (keeping competitors away by the threat of a patent infringement action) and to make available much of the underpinning knowledge. The corollary is that products in the public domain are less likely to be commercialised than those which are privately owned.

Innovations in agriculture, including new plant varieties – and especially plants with advantages that can only be delivered by genetic modification – can contribute to improved food security for smallholder farmers, but only where patents and plant variety rights can provide a reward to justify and return the cost of investment. This is, however, just the start: a more important question is not whether patents do contribute, but whether they can do so more effectively.

111

Unlike real or personal property, which has some intrinsic, physical boundaries, the boundaries of intellectual property are not at all clear

Intellectual property has some interesting characteristics – of which a main one is that the specific rules are social constructs with limited underlying moral or physical principles. Unlike real or personal property, which has some intrinsic, physical boundaries (if I own a car the extent of the property itself, and the extent of my rights over the property, are fairly clear), the boundaries of intellectual property are not at all clear. Of course many of the rules have a long history, many are reflected in treaties that are respected internationally, and within broad limits we can accurately describe the rules of, say, patents or plant variety rights. But only within broad limits, because decade by decade the details of the rules change, by treaty revision, by new treaties, by patent office policies and guidelines, and by judgements in legal cases. For example, the duration of intellectual property rights is often changing, usually becoming longer.

To put it another way, the rules and principles of intellectual property are dynamic, responding to commercial and technical need – for example, early patents granted in a new field of research tend to be broader, reflecting the greater risk of the inventor in exploring new avenues, whereas later patents in a more mature field tend to be narrower. Similarly there needs to be adjustment to the rules to maintain a commercial balance between patent owners and the public, so that the patents don't confer an exclusivity that is too wide and provides excessive reward, or too narrow and fails to justify the cost of research.

Which is where countries and stakeholders can play their part in working with the patent system to ensure that the *precise* rules of intellectual property properly meet the needs of the country and (with reference to the theme of this essay) of smallholder farmers. That means understanding very clearly how patent rules affect them, whether the balance between patent owners and patent users (or society as whole) is about right, and what needs to change. It means engaging in treaty revisions and negotiating for exceptions and derogations, and for changes to the patent rules so that there is proper benefit on both sides – which is the essence of the patent system, and one of its strengths: that it is a dynamic system capable of adapting to changing technical and commercial need.

It is not a trivial nor an easy process, of course, and one in which the developed countries have a strong history and long lead. There are examples of success for both sides: the World Trade Organization's Agreement on Trade-Related Aspects of Intellectual Property Rights (TRIPS), for example, arguably places an inappropriate set of obligations on some developing countries too early in their economic progress, when they would have benefited from a reduced set of intellectual property rules which expanded as commercial and research resources developed. On the other hand the balance in PVR compared to patents is much more towards open use and freedom by breeders (inventors) – what is protected is the variety itself, and not the underlying germplasm, and it is an express feature of PVR that protected varieties can be used freely to breed new varieties (which will then be owned by the breeder of the new variety). Nevertheless, PVR has been adapted by some countries to allow much greater freedom for farmers to save and use their own seed – to the detriment of breeders but the

... the patent system needs to operate efficiently – both the precise rules under which patents are granted and rights exercised, and the effectiveness of the rule of law to give those rights teeth

113

benefit of farmers, which is a sensible response to the role of farmers in many countries in distributing seed. On the whole the recognition of the important work of farmers and their involvement in plant breeding is developing slowly.

Which comes back to the start: patents don't hold up progress, and can improve progress, but the patent system needs to operate efficiently – both the precise rules under which patents are granted and rights exercised, and the effectiveness of the rule of law to give those rights teeth and therefore meaning: where a patent is infringed, there must be an effective legal system to judge the case and enforce the judgement. Governments and other stakeholders need to participate actively in the ongoing development of patents and other forms of intellectual property, to ensure that the best interests of their countries and farmers are being served.

Innovation – and progress in food security – will both flourish when companies are able to justify the cost of R&D by the rewards that patenting can provide, for which the system overall needs to be attractive enough, and reliable enough – enough, but no more.

Dr Sean Butler *is a Fellow of St Edmund's College, Cambridge, where he teaches the law of Ancient Rome, and Intellectual Property.*
St Edmund's College, Cambridge CB3 0BN, UK. scb46@cam.ac.uk

Genetically modified crops: a moral imperative?

McPhoto/Blickwinkel/Specialist Stock

Jürgen Mittelstrass

ased on the achievements of science and technology, our world is becoming more and more the product of humankind. This product includes nature. Natural structures recede and artificial structures take precedence. And with this, many of the problems that we have with ourselves and with our world are also on the rise – especially environmental problems that are characterised by the problematical interweaving of natural and artificial developments, that is, technological developments. The investigation of these developments, insofar as it involves research on the earth and its climate, already presents a difficult scientific task. As formulated by the Max Planck Society in Germany, it includes the investigation of spatial and temporal variations in

... the expected addition of more than 2 billion people ... together with the predicted impact of climate change, will have dramatic consequences

115

*... nature is always creative nature (*natura naturans*) or created nature (*natura naturata*)*. *The first is the nature we cannot live without; the other is the nature that we increasingly appropriate*

structures and in composition of all terrestrial systems from the inner core to the outer atmosphere; the investigation of the connections between physical and chemical processes, which takes account of the energy transfer between the components of the earth–sun system; the investigation of marine and terrestrial ecosystems and their evolution; and the interactions of the biosphere with the processes of the 'system of the earth' – and on top of that must also consider the feedback between the physical, chemical, geological, biological and social systems of the earth, their development and their effects on the metabolism of organisms and the biological complexity of the planet earth.[1]

An important part of this task – from the point of view not only of research but also of intervention – is presented by agriculture. Here we are dealing with feeding a rapidly growing world population and with the sheer survival of our poorest people, who are suffering the consequences of the climate and of difficult economic structures, for instance, in Africa. About 1 billion of the world's population of some 7 billion people, among them one-third of the population of Africa (about 1 billion people), are currently undernourished; and the expected addition of more than 2 billion people within the next two to four decades, together with the predicted impact of climate change, will have dramatic consequences. Above all, population growth leads to a situation in which ever more people in Africa have to live on the product of ever smaller areas of land, which climate-dependently provide ever smaller harvests due to drought, soil degradation, erosion and water scarcity.

In this situation, genetic modification technology, that is, the production of genetically modified food plants, has come into the foreground because of its potential to enhance calorific and nutritional quality (as in the case of pro-vitamin-A-fortified 'Golden Rice' which is pertinent to the treatment of childhood blindness) and to increase resistance of plants to pests and diseases, as well as improving tolerance to environmental stress.

A report from the Pontifical Academy of Sciences provides examples of how transgenic plants could contribute to food security in the context of development.[2] Here, the often-heard objection that by messing around with nature, genetic engineering means 'playing God' is fundamentally misleading. It overlooks the fact that the genetic technologies of plant breeding are merely the continuation of breeding techniques which have been pursued for thousands of years – now with different, more effective means. Furthermore, the fact is usually overlooked that nature, too, is experimenting every day and every night. Some effects cannot be foreseen, as in the emergence of new diseases, but others are a feature of evolution and entirely predictable, as in the appearance of resistance to chemical intervention in medicine and agriculture. They are part of life and nature. This also means that the possible evolutionary risks of genetic engineering are not greater than the risks of the natural process of biological evolution. This is already stated in a report published in 1989 by the US National Research Council (NRC): '*As the molecular methods are more specific, users of these methods will be more certain about the traits they introduce into the plants and hence less liable to produce untoward effects than other methods of plant breeding.*'

Generally speaking, without the application of new technologies as part of the galaxy of other essential changes required in terms of practices, inputs and policies, we will not be able to cope with hunger on this earth, with hunger in

Africa, a continent that struggles not only with climate problems but also with difficult political and social circumstances. In light of this, such application is not only possible and scientifically and economically suggested, but also imperative. Or as the above-mentioned study reports: *'There is a moral imperative to make the benefits of genetic engineering technology available on a larger scale to poor and vulnerable populations who want them and on terms that will enable them to raise the standard of living, improve their health and protect their environments'.*[2] In this sense, particularly if even mere survival depends on the application of these and other technologies coupled with access to fair markets, infrastructure, communications, etc. (which are beyond the scope of this essay), these new technologies must be seen as a global public good – just as hunger is a public challenge. And we should not make a business of hunger. That means that measures must be taken to provide poor farmers in the developing world, as in Africa, with improved crops produced by advanced conventional breeding and, where appropriate, with access to improved genetically engineered crop varieties that fit the local conditions, and to push governments and international aid agencies to invest further in this enterprise. This holds especially for international organisations like the FAO (Food and Agriculture Organization of the United Nations), UNDP (United Nations Development Programme), CGIAR (Consultative Group on International Agricultural Research) and UNESCO (United Nations Educational, Scientific and Cultural Organization), which all bear a special responsibility for the nourishment and physical well-being of the world population.

To return to my starting point: nature is always creative nature (*natura naturans*) or created nature (*natura naturata*). The first is the nature we cannot live without; the other is the nature that we increasingly appropriate, that we seek to optimise for justified purposes, but that we also destroy both knowingly and unwittingly.

With regard to the first sense (creative nature) the point in the future will be to give back to nature a piece of independence, to view it once again more in the sense of the ancient Greek philosopher, Aristotle. This task is supported by ecological points of view and measures. With regard to the second sense (created nature) the point will be to adapt natural developments to the solution of urgent problems of nourishment. This task is served by the new genetic technologies of plant breeding. What is important is that this is done judiciously so as to grant nature its own value, so that nature does not completely become an artifact and at the same time can provide that basis for life that is urgently needed by a growing world population, especially its poor and poorest – and thus most vulnerable – parts.

References
1   www.mpg.de/forschungsgebiete/CPT/GEO/Geo_Klimaforschung/index.html (27.07.2007).
2   Pontifical Academy of Sciences (2010) Report of Conference, Vatican City, 15–19 May. *New Biotechnology* **27**: 445–718.

Professor Dr Jürgen Mittelstrass is at the Department of Philosophy, University of Konstanz, Germany, and is Chairman of the Austrian Science Council (Vienna). Fachbereich Philosophie, Universität Konstanz, D-78457 Konstanz, Germany. juergen.mittelstrass@uni-konstanz.de

Postscript

Adam Rutherford's blog in the UK's *Guardian* newspaper[1] speaks of how scientists responsible for many of the major scientific advances of the modern era are accused of playing God. *'It's not exactly clear what "playing God" actually means,'* he says. *'If playing God involves developing technologies that cure diseases, clean up pollution and create new forms of fuel, then these potential benefits need to be considered without the burden of vague, simplistic soundbites.'*

This restless quest for knowledge is nothing new. It means that science never stands still or can claim to be complete. Therefore this collection of essays is but a small part of what is happening in Africa, namely, African scientists working in African laboratories and using biosciences for African farming, and in particular for African smallholder farmers.

An impressive example is the International Institute for Tropical Agriculture (IITA) with headquarters in Ibadan, Nigeria and multiple centres in 15 African countries.[2] As a member of the Consultative Group on International Agricultural

Research (CGIAR), it is a partner in the Cowpea Productivity Improvement Project, a public–private partnership project that brings together a variety of public, private, research, funding and advocacy organisations.[3]

Plans are currently under way to backcross a selected resistant cowpea strain into a local variety, but further developments depend on the adoption of a National Biosafety Law. If successful the Nigerian cowpea experiment could be used as a model for other West African nations, as in Burkina Faso and Ghana.[4] Nigeria is the largest producer and consumer of cowpea in the world; about 5 million of the total 12.76 million hectares of land are devoted to cowpea, which everywhere is grown mainly by women on small plots of land. It is the most important indigenous African legume in Nigeria because of its ability to grow in drought-prone areas and improve soil fertility. Cowpea pod borer (*Maruca vitrata*) is particularly detrimental to the success of the crop, with reports of crop yield losses as high as 70–80 per cent; overall, insecticides have been ineffective.

In Ghana, the confined field trials (CFTs) of genetically modified rice, cowpea and sweet potato are expected to last for at least three years. They will allow scientists to critically analyse seed quality and ensure the desired traits have been successfully introduced before being recommended for commercial production. The development of GM sweet potatoes aims to combat malnutrition in rural areas by increasing their content of essential amino acids, while GM rice varieties are needed to improve tolerance to drought and salinity in fields that have accumulated salts over the years, forcing farmers to abandon the land.[5]

Journalists provide an all-important link between the discoveries of scientists and their uptake into farming practice. Local African journalists who are B4FA.org Media Fellows[6] write of the encouraging expansion of plant breeding within the continent.

Government involvement

East African highland cooking bananas are a unique type of starchy banana that for centuries has been a major staple for millions of people in the Great Lakes region of East Africa. The crop suffers many productivity constraints that have led the national research systems to adopt advanced breeding techniques in order to provide resistant varieties for poor farmers. Scientists from Uganda and Kenya are engaged in an ongoing experiment in which green pepper genes, a gift from Academica Sinica, Taiwan, have been introduced into local bananas to enable them to resist the banana bacterial wilt (BBW) disease which is fast killing the crop and destroying livelihoods.[7] How this occurs is the subject of intensive research in public and private-sector laboratories as there is currently no commercial chemical, biocontrol agent or any resistant variety that could control the spread of BBW. Dr Andrew Kiggundu of the National Agricultural Research Laboratories, Kawanda points out that even if some BBW-resistant GM varieties were available soon, the legal requirements to guide implementation of the National Biotechnology and Biosafety Policy have yet to be passed. There is a growing realisation that governments play a key part in ensuring both safety and support for the future commercialisation of GM bananas and other African crops that have been bred to resist chronic diseases.

Michael J. Ssali in Uganda's *Saturday Monitor* (24 November 2012) speaks of the country's challenges – rapidly growing population, smaller plots due to fragmentation, exhausted soil mainly because of bad farming practices, farmers who lack the knowledge and financial capacity to apply modern methods to increase yields, and a small national budget for agriculture. But he reports how Ugandan researchers are hard at work to find ways to protect the crops that feed so much of their population. Conventional plant breeding cannot always solve the problems so biotechnology and GM crops have a role to play, particularly for crops as challenging to breed as cultivated bananas, which do not produce seeds and have to be propagated by cuttings.[8]

Even so, once one breakthrough has been achieved, it is a truism that another problem is often around the corner. Dr Africano Kangire of the Coffee Research Centre at Kituuza, Uganda, speaks of coffee wilt disease-resistant varieties which are now available through conventional plant breeding techniques. But then a new problem has surfaced – the black coffee twig borer – a new challenge that is fast reducing coffee yields.[8] So the task of the scientist is to keep one step ahead of the game.

Henry Lutaaya, a B4FA Media Fellow who writes for *Sunrise* in Uganda, says that increasing his knowledge and awareness about plant breeding was one method for beginning to take action about Uganda's food security: '*Journalists had the opportunity to learn about plant breeding concepts such as hybridization, tissue culture, genetically modified organisms.*' One of the most interesting aspects

International collaboration

A theme throughout the essays is the success of international collaboration.[9] Over 50 per cent of the world's cassava production occurs in Africa where cassava is used not only as food, feed and beverage but for products such as paper, wood, textiles and biofuels. The goal is to improve cassava's resistance to the viral diseases, cassava brown streak disease (CBSD) and cassava mosaic disease (CMD). This involves testing resistant varieties in the field for their stability against disease, and obtaining regulatory approval for making them available to smallholder farmers. The Virus Resistant Cassava for Africa (VIRCA) project engages the National Crops Resources Research Institute, Uganda, the Kenya Agricultural Research Institute (KARI), Kenya, and the Donald Danforth Plant Science Center, USA. Results under greenhouse conditions have been promising, as also have confined field trials in East Africa. The long-term aim is to deliver royalty-free improved planting materials for farmers. Scientist Dr Douglas Miano of KARI also emphasises that the VIRCA project helps to build the capability of local scientists in crop biotechnology; it is an application to an African crop by African scientists.

was the discovery that Uganda is itself a leader in plant breeding and bio-technology research – and the answers to Uganda's food security may very well lie in the hands of its own talented crop of researchers.[10]

In conclusion, we can look back to how farming emerged between hunting–gathering and settled agriculture about 10,000 years ago.[11] How our ancestors started to till the fields, plant the first seeds and select the best plants which began to change the genetic composition of plant populations. How this led to a gradual divergence between wild and cultivated members of plant species.

But in these essays we fast-forward to brilliant scientists of the present century who hunt for genes and gather knowledge about plant breeding – what information genes carry, what they do, how they can be selected, how they are controlled, and how they can be marshalled to help the challenges of a planet threatened by overpopulation, overconsumption and unequal distribution of its wealth and resources.

But there is a difference between acquiring knowledge and information and possessing wisdom,[12] and there is a proverb which says: *'Wisdom is supreme, therefore get wisdom. Though it cost you all you have, get understanding.'*[13] As science, technology and innovation bring to society their successes and rewards, they demand to be handled wisely[14] with structures of oversight and transparency of communication that ensure they are sustainable, safe and acceptable for consumers – all of us!

These essays are central to the B4FA project (www.b4fa.org) which has been designed to enlarge understanding of how the biosciences can help farming and food security in Africa. It seeks to capture the idea about which Nobel Laureate, Sydney Brenner, wrote recently: *'The whole of biology must be rooted in*

DNA, and our task is to discover how these DNA sequences arose in evolution and how they are interpreted to build the diversity of the living world. Physics was once called natural philosophy; perhaps we should call biology "natural engineering".[15]

Brian Heap   **David Bennett**

References

1   Rutherford A. (2012) Synthetic biology: 'playing God' is vital if we are to create a better future for all. *Guardian*, 27 July 2012. www.guardian.co.uk

2   International Institute for Tropical Agriculture (2012) *Annual Report 2011*. www.iita.org/annual-reports

3   www.agricultureandfoodsecurity.com/content/1/S1/S5

4   www.aatf-africa.org/userfiles/cowpea-2011-progress-report.pdf

5   www.aatf-africa.org/cowpea/news/ghana_to_begin_gmo_testing/en

6   Biosciences for Farming in Africa: catalyzing informed discussion about biosciences and plant genetics for farmers in Africa. www.b4fa.org

7   www.scidev.net/en/news/gm-banana-resistant-to-fungus-shows-promise-1.html; http://www.iita.org/bananaplantain-asset/-/asset_publisher/9zYD/content/green-pepper-to-the-rescue-of-african-bananas?redirect=%2Fbanana-and-plaintain

8   www.monitor.co.ug/Magazines/Farming/Biotechnology-has-the-potential-to-solve-our-farming-problems/-/689860/1613274/-/so3xnkz/-/index.html

9   www.slideshare.net/AfriCenter/dr-douglas-miano-overview-of-the-virus-resistant-cassava-virca-project#btnNext

10 www.sunrise.ug/news/top-stories/4680-b4fa-fires-up-journalists-with-knowledge-on-modern-plant-breeding.html

11 Kingsbury N. (2011) *Hybrid: The History and Science of Plant Breeding*. Chicago, IL: University of Chicago Press, pp. 16–17.

12 Templeton J.M. (1997) *Laws of Life*. Philadelphia, PA: Templeton Foundation Press, p. 28.

13 Proverbs 4:7.

14 Academy of Science of South Africa (2012) *Regulation of Agricultural GM Technology in Africa*. www.assaf.org.za

15 Brenner S. (2012) History of science: the revolution in the life sciences. *Science* **338** 1427–8.

Contributors

Professor Walter S. Alhassan is a consultant on biotechnology and biosafety policy issues for the Forum for Agricultural Research in Africa (FARA) in Accra, Ghana. He was the Coordinator of the Program for Biosafety Systems (PBS) for West Africa, a founding member and later Chairman of the Board of Trustees of the African Agricultural Technology Foundation (AATF), and Director-General of the Ghana Council for Scientific and Industrial Research (CSIR). He is a Fellow of the Ghana Academy of Arts and Sciences (GAAS) and Member of the Ghana Atomic Energy Commission.

Professor Sir David Baulcombe is the Regius Professor of Botany and Royal Society Research Professor at the Department of Plant Sciences of the University of Cambridge. He is a molecular biologist and his research interest in plants focuses on how genes can be silenced, and how nurture can influence nature. This research links to disease resistance in plants and understanding of hybrids. Extramural activities include membership of the Biotechnology and Biological Sciences Research Council and in 2009 he chaired a Royal Society policy study on the contribution of biological science to food crop productivity – *Reaping the Benefits: Science and the Sustainable Intensification of Global Agriculture.*

Dr David Bennett is co-Project Leader, *Biosciences for Farming in Africa* (B4FA.org), Visitor to the Senior Combination Room of St Edmund's College, Cambridge, UK and Guest at the Department of Biotechnology of the Delft University of Technology in The Netherlands. He has a PhD in biochemical genetics and an MA in science policy studies with long-term experience of the relations between science, industry, government, education, law, the public and the media. David works with the European Commission, government departments, companies, universities, public-interest organisations and the media in the UK, USA, Australia and The Netherlands. He is co-editor of *Successful Agricultural Innovation in Emerging Economies: New Genetic Technologies for Global Food Production* (Cambridge University Press, 2013).

Phil Bloomer is Director of the Campaigns and Policy Division at Oxfam GB. He is responsible for a team of 170 staff working across policy, advocacy, programme advice, campaigns and social media. Current global priorities are: the GROW campaign (food justice in a resource-constrained world) which focuses on food, climate, land and water issues; humanitarian protection and assistance in crises; and the provision of essential health and education for all. Phil is a member of the Global Food Strategy Advisory Board and the Foresight Global Food and Farming Futures High Level Stakeholder Group. Before joining Oxfam, he worked for 11 years in Latin America on international economic justice issues, human rights and humanitarian emergencies.

Dr Sean Butler is a Fellow of St Edmund's College, Cambridge, where he teaches the law of Ancient Rome, and Intellectual Property. His fields of interest include intellectual property, knowledge transfer from the public sector, and business strategy. He has worked in developing countries over many years advising Centres of the CGIAR on intellectual property, impact, and delivery of results to stakeholder farmers. He is also Head of Strategy at NIAB (National Institute of Agricultural Botany) in Cambridge.

Professor Eric Yirenkyi Danquah is Professor of Plant Molecular Genetics in the Department of Crop Science, University of Ghana. He is the founding Director of the West Africa Centre for Crop Improvement (WACCI), which trains a new generation of plant breeders for the West African sub-region. He has a BSc degree (crop science) from the University of Ghana, and MPhil (plant breeding) and PhD (genetics) degrees from the University of Cambridge. His research focuses on using the tools of genomics to facilitate crop improvement and is underpinned by an understanding of how genetic diversity in crop plants and their associated pests relates to performance.

Dr Roger Day is Deputy Director, Development, at the Commonwealth Agricultural Bureau International (CABI) regional centre in Nairobi. He has 30 years' experience of tropical agricultural research and development in Africa and Asia, 20 of them with CABI in Africa, including three as Regional Director. His work is concerned with the application of science and research for development. His interests span extension, communication and capacity development, with a technical focus on plant health in relation to production and trade.

Dr Joe DeVries is Director of the Program for Africa's Seed Systems (PASS), Alliance for a Green Revolution in Africa (AGRA) based in Nairobi, Kenya. This is a US$200 million initiative operating in 17 African countries which has trained 122 PhD students and 202 MSc students in plant breeding and crop sciences, achieved the breeding of nearly 400 new crop varieties, assisted in the creation of over 70 private, independent, African seed companies, and increased Africa's certified seed supply by over 40,000 million tons annually. Joe has lived and worked in Africa since 1986, focusing on increasing the productivity of Africa's farmers. Following his MSc in crop–water relations he served as a United Nations volunteer and NGO programme manager in northern Mali building irrigated rice schemes among villagers, and managed a large emergency seed supply programme in Mozambique. He has a PhD in genetics and breeding from Cornell University, worked with World Vision International as Director for Food Security in Africa, and spent 12 years with the The Rockefeller Foundation, where he led a team to create what became AGRA's PASS programme.

Dr Marco Ferroni is Executive Director, Syngenta Foundation for Sustainable Agriculture, Basel, Switzerland. He is an expert on international agriculture and sustainability issues. Marco joined the Syngenta Foundation after a career in multilateral institutions and government. Before joining the Foundation, he worked at the Inter-American Development Bank and the World Bank in managerial and advisory roles. Earlier in his career he was an economist and programme director for economic affairs, international trade and development cooperation in the government of Switzerland. Marco Ferroni holds a PhD degree in agricultural economics from Cornell University.

Professor Sir Brian Heap FRS is Project Leader, *Biosciences for Farming in Africa* (B4FA.org), Research Associate, Centre for Development Studies, University of Cambridge, and President of the European Academies Science Advisory Council. He was Master of St Edmund's College, Cambridge, and Vice-President and Foreign Secretary of the Royal Society. He has doctorates from Nottingham and Cambridge and was Director of Research at the Institute of Animal Physiology and Genetics Research (Cambridge and Edinburgh) and at the Biotechnology and Biological Sciences Research Council (Swindon). He has been engaged in public issues of biotechnology, population growth, sustainability and science policy working with the World Health Organization, the UK–China Forum and the European Commission.

Professor Calestous Juma FRS is Professor of the Practice of International Development and Director of the Science, Technology, and Globalisation Project at Harvard University. He directs the Agricultural Innovation in Africa Project funded by the Bill & Melinda Gates Foundation and serves as Faculty Chair of Innovation for the Economic Development executive programme. Calestous is former executive secretary of the UN Convention on Biological Diversity and co-chair of the African Union's high-level panel on science, technology and innovation. He has been elected to several scientific academies including the Royal Society of London, the US National Academy of Sciences, the Academy of Sciences for the Developing World, the UK Royal Academy of Engineering and the African Academy of Sciences. He has won several international awards for his work on sustainable development. He holds a doctorate in science and technology policy studies and has written widely on science, technology and the environment. He serves on the jury of the Queen Elizabeth Prize for engineering and is writing a book on resistance to new technologies.

Dr Daniel Karanja works as a plant pathologist at CABI in Africa. He is a career scientist with a PhD in agriculture. Daniel has broad technical knowledge in all aspects of seed health, the development of sustainable seed systems through public–private partnership and value chains in developing countries. He was part of a scientific team that led to the participatory development and release of two locally developed varieties of kale for commercialisation in Kenya. He now leads an innovative initiative on scaling up economically viable and sustainable farmer-led seed enterprises to enhance smallholder farmers' access to quality seeds of African indigenous vegetables, and boost household incomes and nutrition in Eastern and Central Africa.

Dr Margaret Karembu is Director of the International Service for the Acquisition of Agri-biotech Applications (ISAAA) AfriCenter, based in Nairobi, Kenya. She is a senior environmental science management specialist with extensive experience in science communication and agri-biotechnology applications in Africa. Margaret has been actively involved in strengthening capacity for biosafety communications and policy outreach for informed choices on the acquisition and transfer of modern agricultural biotechnology by small-scale farmers in sub-Saharan Africa. She holds a PhD degree in environmental science education from Kenyatta University, Kenya and is chairperson of the Co-operative University College of Kenya Council, a constituent of the Jomo Kenyatta University of Agriculture and Technology (JKUAT).

Dr Segenet Kelemu is Director of the Biosciences eastern and central Africa (BecA) Hub, Nairobi, Kenya. She grew up in a remote village in Ethiopia, bearing the unequal burden carried by rural African women. Her drive for success and achievement despite any obstacle was ingrained from early on and she has become an influential molecular plant scientist, contributing several novel products for use in global agriculture. She joined the International Center for Tropical Agriculture (CIAT) as a Senior Scientist and was later appointed as its Leader of Crop and Agroecosystem Health Management. As Director of the BecA Hub she has been influential in transforming it from a contentious idea into a driving force that is changing the face of African biosciences. She received the Outstanding Senior Scientist Award at CIAT, the prestigious Friendship Award granted by the People's Republic of China and the Academy of Sciences for the Developing World (TWAS) Prize for Agricultural Sciences.

Dr Denis Kyetere is Executive Director of the African Agricultural Technology Foundation (AATF). Previously he was Director-General of Uganda's National Agricultural Research Organisation (NARO). He began his career at NARO as a scientific officer and maize agronomist and later as head of the maize programme, and programme leader for cereals, among others. He led a team that identified and mapped the maize streak virus gene 1 (MSV1) which confers tolerance in maize to MSV disease, and also developed several maize varieties that are grown in Uganda and surrounding countries. He was the first chairperson of the Executive Advisory Board of AATF's Water Efficient Maize for Africa (WEMA) project. Denis has a PhD from Ohio State University (OSU), an MSc from the University of Wales, Aberystwyth College, UK – all in genetics and plant breeding – and a BSc in botany and zoology from Makerere University, Uganda.

Dr Julian Little is UK Communications and Government Affairs Manager of Bayer CropScience and has worked in plant science and crop production for over 20 years. With a PhD in molecular plant pathology from the University of Wales, he joined Rhone-Poulenc as a plant biochemist working in both the UK and France, and then in research project management in Aventis CropScience. He is based in Cambridge dealing with media, political and public enquiries on the subject of crop production, innovative plant breeding, and sustainable agriculture. His role also covers policy and corporate social responsibility. Julian is Chair of the Agricultural Biotechnology Council (ABC), representing the major companies interested in the application of biotechnology in crops. He also chairs the respective communications groups of the Crop Protection Association in the UK and EuropaBio in Brussels.

Dr Niels Louwaars is Director of the Dutch seed association Plantum and is also member of the the Law and Governance Group at Wageningen University, The Netherlands. He was trained as a plant breeder and worked in seed programmes in Asia and Africa before concentrating at Wageningen University on seed policies and regulations, including rights on genetic resources, traditional knowledge and intellectual property. He introduced the concept of 'integrated seed systems', putting formal and local seed systems in one frame. His PhD focused on the combined impact of various international agreements affecting the diversity of seed systems that farmers use.

Professor Dr Jürgen Mittelstrass is at the Department of Philosophy, University of Konstanz, Germany, and is Chairman of the Austrian Science Council (Vienna). He studied philosophy, German literature and protestant theology at the universities of Bonn, Erlangen, Hamburg and Oxford. He has a PhD in philosophy from the University of Erlangen and became Professor of Philosophy and Philosophy of Science at the University of Konstanz. He was President of the German Philosophical Association and President of the Academia Europaea (London). In 1989 he was awarded the prestigious Leibniz Prize of the Deutsche Forschungsgemeinschaft (DFG). His many publications include *Mind, Brain, Behavior* (1991) and *Leibniz und Kant* (2011), and he is editor of *Enzyklopädie Philosophie und Wissenschaftstheorie*, 4 vols. (1980–1996, 2nd edn in 8 vols., 2005ff.).

Nick Moon MBA FRSA is Executive Chairman of Wanda Organic (www.wandaorganic. com), a start-up for-profit social enterprise offering the latest soil and plant health solutions to African farmers, and is a voluntary Director of Peace for Africa and Economic Development (www.padinst.org), a youth-focused peace-building and job-creation initiative. He has lived in Kenya since 1982, building up 30 years of experience working in the social and economic development sector. In 1991 he co-founded and, until December 2011, was the Managing Director of the award-winning not-for-profit social enterprise KickStart International (www.kickstart.org), recently listed by *Forbes* Magazine as one of the world's Top 30 Social Enterprises, and by the *Global Journal* as one of the Top 100 NGOs. Nick has now handed over all responsibilities for its operational management in Africa, but continues as a member of the Board of Directors and represents the organisation at high-level conferences and forums.

Dr Sylvester O. Oikeh is Project Manager of the Water Efficient Maize for Africa (WEMA) project, a position he has held since 2009. He has more than 18 years' experience in research and development projects on natural resources and crop management, and plant nutrition. He joined the African Agriculture Technology Foundation (AATF) from the Africa Rice Centre, Benin, where he served as a Principal Scientist/Soil Fertility Agronomist and Project Leader. Sylvester has also worked at the International Institute of Tropical Agriculture (IITA). He was a postdoctoral fellow at Cornell University's US Department of Agriculture (USDA) Plant, Soil and Nutrition Laboratory at Ithaca, New York. Sylvester attained his doctorate in soil science from the Ahmadu Bello University, Nigeria, an MSc in crop science from the University of Nigeria, Nssuka, and a Bachelor of Agriculture degree in horticulture from the same university.

Professor Sir Ghillean Prance FRS is Scientific Director and a Trustee of the Eden Project in Cornwall and Visiting Professor at Reading University, UK. He is active in plant systematics and in conservation of the tropical rainforest. He obtained a BA in botany and a DPhil at Keble College, Oxford. He began at the New York Botanical Garden in 1963 as a research assistant and subsequently B.A. Krukoff Curator of Amazonian Botany, Director and Vice-President of Research and finally Senior Vice-President for Science. His exploration of Amazonia included 15 expeditions in which he collected over 350 new species of plants. He was Director of the Royal Botanic Gardens, Kew, UK, McBryde Professor at the National Tropical Botanical Garden in Hawaii and is currently McBryde Senior Fellow there. He is author of 19 books and has published more than 520 scientific and general papers in taxonomy, ethnobotany, economic botany, conservation and ecology. He holds 15 honorary doctorates, has received the International Cosmos Prize, has been elected a Fellow of the Royal Society and holds many other awards.

Dr Dannie Romney is Global Director for Knowledge for Development at CABI, Nairobi, Kenya. It is one of four themes that support activities designed to get research into use in Africa, South and South East Asia as well as in Latin America. Work under this theme includes activities to develop and strengthen seed systems for a variety of crops including cereals, vegetables and African indigenous vegetables. Dannie has lived in Africa for 13 years and worked with smallholder farming systems in Africa, Asia and Latin America for 24 years. Before joining CABI she worked for seven years at the International Livestock Research Institute (ILRI), looking at nutrient recycling and feed resources with a focus on smallholder intensive dairy mixed crop–livestock systems. She was Acting Director for the ILRI Innovation Systems theme.

Professor Jennifer A. Thomson is Emeritus Professor in the Department of Molecular and Cell Biology at the University of Cape Town. Jennifer has a BSc in zoology from the University of Cape Town, an MA in genetics from Cambridge and a PhD in microbiology from Rhodes University. She was a postdoctoral fellow at Harvard Medical School, Associate Professor at the University of the Witwatersrand and Director of the Council for Scientific and Industrial Research (CSIR) Laboratory for Molecular and Cell Biology before becoming Professor and Head of the Department of Microbiology at the University of Cape Town. Her main research interests are the development of maize resistant to viruses and tolerant to drought. She has published two books: *Genes for Africa: Genetically Modified Crops in the Developing World* and *GM Crops: The Impact and the Potential*.

Grace Wachoro is the Corporate Communications Officer at the African Agricultural Technology Foundation (AATF). She is responsible for supporting overall AATF communication strategy development, implementation and monitoring. Prior to her current position, Grace was the Project Communications Officer for the Water Efficient Maize for Africa (WEMA) project from 2008 to 2011. She holds a Master's degree in gender and development and a postgraduate diploma in mass communication, both from the University of Nairobi, and a BSc degree in information sciences from Moi University, Kenya. Prior to joining AATF, she worked for the Institute of Policy Analysis and Research, Nairobi, as an editor. She previously worked for various organisations as an editor and writer.

Dr Florence Wambugu is a renowned African scientist who is currently the CEO and Founder of Africa Harvest Biotech Foundation International (Africa Harvest, www.africaharvest.org). She is also the Project Co-Principal Investigator of the Africa Biofortified Sorghum (ABS) Project, which was initially funded by the Bill & Melinda Gates Foundation and is now funded by the Howard G. Buffett Foundation through DuPont-Pioneer. The project is made of 13 public–private member institutions and a network of over 70 scientists in Africa and the USA in a 'virtual institution'. The objective of the project is to enhance sorghum nutrition with vitamin A, which has the potential to benefit close to 300 million African consumers. Florence gained her BSc degree in botany from the University of Nairobi, Kenya, her MSc in pathology from North Dakota State University, USA and her PhD in virology and biotechnology at the University of Bath, UK. In 1992–1994 she worked as a postdoctoral research associate at Plant Science Monsanto, St Louis, USA. In 2009 she was awarded an honorary DSc degree from the University of Bath, UK.

The editors are deeply indebted to the Team and Advisers of B4FA (b4fa.org) and many colleagues at home and abroad for their help and advice in the preparation of these essays.

Index